IMMUNODEFICIENT RODENTS

A Guide to Their Immunobiology, Husbandry, and Use

Committee on Immunologically Compromised Rodents
Institute of Laboratory Animal Resources
Commission on Life Sciences
National Research Council

NATIONAL ACADEMY PRESS
Washington, D.C.   1989

NATIONAL ACADEMY PRESS • 2101 Constitution Avenue, NW • Washington, DC 20418

NOTICE: The project that is the subject of this report was approved by the Governing Board of the National Research Council, whose members are drawn from the councils of the National Academy of Sciences, National Academy of Engineering, and Institute of Medicine. The members of the committee responsible for the report were chosen for their special competences and with regard for appropriate balance.

This report has been reviewed by a group other than the authors according to procedures approved by a Report Review Committee consisting of members of the National Academy of Sciences, National Academy of Engineering, and Institute of Medicine.

The National Academy of Sciences is a private, nonprofit, self-perpetuating society of distinguished scholars engaged in scientific and engineering research, dedicated to the furtherance of science and technology and to their use for the general welfare. Upon the authority of the charter granted to it by the Congress in 1863, the Academy has a mandate that requires it to advise the federal government on scientific and technical matters. Dr. Frank Press is president of the National Academy of Sciences.

The National Academy of Engineering was established in 1964, under the charter of the National Academy of Sciences, as a parallel organization of outstanding engineers. It is autonomous in its administration and in the selection of its members, sharing with the National Academy of Sciences the responsibility for advising the federal government. The National Academy of Engineering also sponsors engineering programs aimed at meeting national needs, encourages education and research, and recognizes the superior achievements of engineers. Dr. Robert M. White is president of the National Academy of Engineering.

The Institute of Medicine was established in 1970 by the National Academy of Sciences to secure the services of eminent members of appropriate professions in the examination of policy matters pertaining to the health of the public. The Institute acts under the responsibility given to the National Academy of Sciences by its congressional charter to be an adviser to the federal government and upon its own initiative to identify issues of medical care, research, and education. Dr. Samuel O. Thier is president of the Institute of Medicine.

The National Research Council was established by the National Academy of Sciences in 1916 to associate the broad community of science and technology with the Academy's purposes of furthering knowledge and advising the federal government. Functioning in accordance with general policies determined by the Academy, the Council has become the principal operating agency of both the National Academy of Sciences and National Academy of Engineering in the conduct of their services to the government, the public, and the scientific and engineering communities. The Council is administered jointly by both Academies and the Institute of Medicine. Dr. Frank Press and Dr. Robert M. White are chairman and vice-chairman, respectively, of the National Research Council.

This project was funded primarily by the U.S. Department of Health and Human Services (DHHS) through contract number NO1-CM-57644 with the National Cancer Institute. Additional funding was provided by the National Science Foundation (NSF) through grant number BSR-8716373. Any opinions, findings, and conclusions or recommendations expressed in this publication are those of the committee and do not necessarily reflect the views of DHHS or NSF, nor does the mention of trade names, commercial products, or organizations imply endorsement by the U.S. government.

Library of Congress Catalog Card No. 89-12387

ISBN 0-309-03796-4

Copyright © 1989 by the National Academy of Sciences

Printed in the United States of America

COMMITTEE ON IMMUNOLOGICALLY COMPROMISED RODENTS

Fred W. Quimby (Chairman), Center for Research Animal Resources, New York State College of Veterinary Medicine, Cornell University
Melvin J. Bosma, Institute for Cancer Research, Fox Chase Cancer Center, Philadelphia, Pennsylvania
Robert A. Good, All Childrens Hospital, St. Petersburg, Florida
Carl T. Hansen, Veterinary Resources Branch, Division of Research Services, National Institutes of Health, Bethesda, Maryland
David D. Myers, Research Animal Resources Center, Memorial Sloan-Kettering Cancer Center, New York, New York
Conrad B. Richter, Division of Laboratory Animal Medicine, Duke University Medical Center
John B. Roths, The Jackson Laboratory, Bar Harbor, Maine
Henry H. Wortis, Department of Pathology, Tufts University School of Medicine

Invited Participants

David E. Briles, The University of Alabama at Birmingham
Muriel T. Davisson, The Jackson Laboratory, Bar Harbor, Maine
Gabriel Fernandes, University of Texas Health Science Center, San Antonio
Margaret C. Green, Emeritus, The Jackson Laboratory, Bar Harbor, Maine
Dennis L. Guberski, University of Massachusetts Medical School
Hedi Haddada, Institut de Recherches Scientifiques sur le Cancer, Villejuif, France
Hans J. Hedrich, Central Institute for Laboratory Animal Breeding, Hannover, Federal Republic of Germany
Masahisa Kyogoku, Department of Pathology, Tohoku University School of Medicine, Sendai, Japan
Priscilla W. Lane, The Jackson Laboratory, Bar Harbor, Maine
Jennifer L. Lasky, Department of Pathology, New York University Medical Center
Edward H. Leiter, The Jackson Laboratory, Bar Harbor, Maine
Sandy C. Marks, Jr., University of Massachusetts Medical School
Thomas H. Roderick, The Jackson Laboratory, Bar Harbor, Maine
Leonard D. Shultz, The Jackson Laboratory, Bar Harbor, Maine

G. Jeanette Thorbecke, Department of Pathology, New York University Medical Center

Sara E. Walker, Harry S. Truman Memorial Veterans Hospital, Columbia, Missouri

Staff

Dorothy D. Greenhouse, Senior Program Officer
Judith Grumstrup-Scott, Consulting Editor

The Institute of Laboratory Animal Resources (ILAR) was founded in 1952 under the auspices of the National Research Council. Its mission is to provide expert counsel to the federal government, the biomedical research community, and the public on the scientific, technological, and ethical use of laboratory animals within the context of the interests and mission of the National Academy of Sciences. ILAR promotes the high-quality humane care of laboratory animals; the appropriate use of laboratory animals; and the exploration of alternatives in research, testing, and teaching.

INSTITUTE OF LABORATORY ANIMAL RESOURCES COUNCIL

Steven P. Pakes (Chairman), University of Texas Southwestern Medical Center, Dallas
June R. Aprille, Tufts University
Melvin W. Balk, Charles River Laboratories, Inc., Wilmington, Massachusetts
Douglas M. Bowden, University of Washington, Seattle
Thomas J. Gill III, University of Pittsburgh School of Medicine
Alan M. Goldberg, School of Hygiene and Public Health, The Johns Hopkins University
Fred W. Quimby, New York State College of Veterinary Medicine, Cornell University
J. Wesley Robb, School of Medicine, University of Southern California
William H. Stone, Department of Biology, Trinity University

COMMISSION ON LIFE SCIENCES

Bruce M. Alberts (Chairman), University of California, San Francisco
Perry L. Adkisson, Texas A&M University
Francisco J. Ayala, University of California, Irvine
J. Michael Bishop, University of California Medical Center, San Francisco
Freeman J. Dyson, The Institute for Advanced Study, Princeton, New Jersey
Nina V. Fedoroff, Carnegie Institution of Washington, Baltimore, Maryland
Ralph W. F. Hardy, Boyce Thompson Institute for Plant Research, Ithaca, New York
Richard J. Havel, Cardiovascular Research Institute, University of California School of Medicine, San Francisco
Leroy E. Hood, California Institute of Technology
Donald F. Hornig, Harvard University School of Public Health
Ernest G. Jaworski, Monsanto Company, St. Louis, Missouri
Simon A. Levin, Ecosystems Research Center, Cornell University, Ithaca, New York
Harold A. Mooney, Stanford University

Steven P. Pakes, University of Texas Southwestern Medical Center, Dallas
Joseph E. Rall, National Institutes of Health, Bethesda, Maryland
Richard D. Remington, University of Iowa
Paul G. Risser, University of New Mexico, Albuquerque
Richard B. Setlow, Brookhaven National Laboratory, Upton, New York
Torsten N. Wiesel, Rockefeller University

Preface

In 1974 the Institute of Laboratory Animal Resources, National Research Council, convened the Committee on Care and Use of the "Nude" Mouse to prepare guidelines for maintaining, breeding, and rearing mice homozygous for the autosomal recessive mutation "nude." These mice, which have thymic aplasia and a developmental defect in hair growth, are difficult to maintain because of their severely compromised T-cell immunity and, consequently, their lack of resistance to many microbial diseases. With their increasing use as animal models, especially in the fields of immunology, oncology, and infectious diseases, it was recognized that guidelines were needed to ensure the production and maintenance of healthy animals. The committee's 1976 report, *Guide for the Care and Use of the Nude (Thymus-Deficient) Mouse in Biomedical Research*, provided such guidelines.

Since then many immunodeficient rodents have been identified, and the study of these models has increased our understanding of the development and function of the immune system. Concurrently, there has been a broadened awareness of the increased susceptibility of immunodeficient rodents to various infectious agents. New construction materials, shipping containers, and animal-care equipment have helped to protect these animals from disease-producing agents. Many immunodeficient strains are now commercially available in the pathogen-free state and are maintained under rigid quality-assurance programs to guarantee their microbial and genetic status. Each of these innovations, however, places greater pressure on the users of these models to plan in advance for their selection, transportation, housing, and maintenance.

The information contained in this volume is intended to assist investigators in selecting appropriate models for immunologic research. Current knowledge about the maintenance and breeding of these models is also included. The Committee on Immunologically Compromised Rodents has designed this book to be used in conjunction with several National Research Council publications, particularly the *Guide for the Care and Use of Laboratory Animals*, which was prepared by the Institute of Laboratory Animal Resources (ILAR) and published in 1985 by the U.S. Department of Health and Human Services.

The committee extends its appreciation to the contributors of this volume and to the staff of ILAR, especially Dr. Dorothy Greenhouse and Judith Grumstrup-Scott. Their dedication to and support of the committee have made the publication of this document possible.

Fred W. Quimby, *Chairman*
Committee on Immunologically
Compromised Rodents

Contents

1 INTRODUCTION 1
  Immune System Function   2
  Effect of Environmental Factors on Immune Function   11
  General Considerations for Maintaining Immunodeficent
    Rodents   14
  Mutations   15
  Gene Markers and Chromosome Maps   16
  Nomenclature and Sources of Immunodeficient Rodents   17
  General Reading   34

2 HEREDITARY IMMUNODEFICIENCIES 36
  Mice with Single Mutations   36
  Mice with Multiple Mutations   90
  Inbred Strains of Mice   94
  Outbred Mice   122
  Rat Mutants   123
  Inbred Strains of Rats   130
  Guinea Pig Mutants   134
  Hamster Mutants   138

3 INDUCED IMMUNODEFICIENCIES 140
Chemical Inducers 140
Infectious Agents 143
Nutrition 144
Ionizing and Ultraviolet Radiation 144
Biological Inducers 145
Thymectomy 146

4 MAINTENANCE OF RODENTS REQUIRING ISOLATION 148
General Considerations 148
Special Facilities and Equipment 149
Specialized Husbandry 151
Control of Infection 153

5 MATING SYSTEMS FOR MUTANTS 155
Inbreeding 155
Propagation Without Inbreeding 159

6 GENETIC MECHANISMS GOVERNING RESISTANCE OR SUSCEPTIBILITY TO INFECTIOUS DISEASES 161

REFERENCES 165

APPENDIX: HEMATOPOIETIC CELL-SURFACE ANTIGENS MENTIONED IN THE TEXT 213

GLOSSARY 217

INDEX 233

1

Introduction

Each of the rodent strains and mutants described in this volume has some defect in immunity. Some of the deficiencies, like that which occurs in the nude mouse, are caused by single point mutations; others are complex abnormalities involving multiple genes. Some of the mutations are allelic; others, although exhibiting similar pathologic processes, are not. The ability to establish these mutations in inbred strains and to develop congenic lines has greatly enhanced the usefulness of these models and fostered a better understanding of mutant gene effects. New techniques, such as DNA cloning and sequencing, have allowed investigators to define precisely the biochemical defect in several mutations. However, when these animals are used, an understanding of mechanisms for ensuring genetic purity and a knowledge of standardized nomenclature are essential.

The purpose of this volume is to summarize and furnish key references on the genetics, pathophysiology, husbandry, and reproduction of various immunodeficient rodent strains and mutants. To aid the reader in evaluating these models, the organization and function of the mammalian immune system are reviewed, and pertinent information on both gene markers and standardized nomenclature is provided. In addition, the role of environmental factors on normal immunologic functions is discussed.

Several of the mutants considered in this volume require strict isolation from infectious agents, and one chapter (Chapter 4) is devoted to detailing isolation procedures. Special approaches are necessary to propagate some of the mutants, and another chapter (Chapter 5) discusses various mating systems. Finally, the information in this document has been extracted from a body of literature spanning 20 years of research in immunology. Naturally, during that time many synonyms and acronyms have emerged, and the definitions of certain terms have been refined. To assist the reader with these terms, the committee has prepared a table (see the Appendix) organizing the rodent lymphocyte differentiation antigens into currently known cluster designation groups and a glossary, arranged by abbreviation followed by the complete name of each term.

IMMUNE SYSTEM FUNCTION

Host Defense Systems

All mammals live in a sea of microorganisms, which includes bacteria, fungi, viruses, rickettsia, protozoa, and multicellular parasites. Many of these organisms have the capacity to replicate within the animal's body and cause disease; however, they are prevented from establishing an infection by a complex system of interacting innate and adaptive immune mechanisms. The innate defense system includes the integrity of the skin and mucous membranes; mucous secretions; cilia of epithelial cells on mucous membranes, which generate a mucous stream; nonspecific inflammatory processes; phagocytic cells; large granular lymphocytes, which function as natural killer cells; and biologically active molecules, such as histamine, complement, and other acute-phase reactants. Innate defense mechanisms can act alone to prevent the introduction of infectious agents into the body or to survey the tissues for newly arising neoplastic cells that, once recognized, are attacked. These nonspecific mechanisms also interact with specific (adaptive) immune mechanisms by processing foreign material prior to its presentation to lymphoid cells (afferent phase), amplifying relevant clones of lymphoid cells (central phase), and assisting in the delivery of an immune-mediated attack on foreign particles (efferent phase). When an infectious agent successfully breaches the innate defense system, the adaptive immune system is activated to mount an attack that is specific to the invading agent. A mechanism of immunologic memory in the adaptive immune system enhances the protection of the host against subsequent assaults by the same agent.

The Nature of Adaptive Immunity

Through eons of selection, vertebrates have evolved a system of interacting cells and molecular substances that operate in a coordinated fashion to deliver an attack of exquisite specificity. This high degree of specificity is often necessary to direct effector cells and substances against targeted agents without also damaging bystander cells and tissues. However, in order to achieve this degree of specificity, two major obstacles must first be overcome. The first is the requirement for a system that can generate an enormous number of unique recognition sites capable of binding to the myriad of potentially harmful agents that animals encounter on a daily basis. The second obstacle is to confine the action of this system to elements, both cellular and molecular, that are foreign to the host. The adaptive immune system has developed novel and very complex solutions to these problems.

The great complexity involved in generating immunologic specificity affords opportunities for errors to arise in the system. These errors, although relatively rare, are frequently devastating to the host. They may be characterized by the inability of the host to eliminate or neutralize foreign substances (immunodeficiency) or the failure to discriminate between self and nonself (autoimmunity).

The adaptive immune system is composed of various functionally distinct cells and their products. Cells of the lymphoid series play a principal role and are characterized by the presence on the cell surface of distinct glycopeptides that serve as developmental markers and functional receptors. Postnatally, pluripotential hematopoietic stem cells reside in the bone marrow. Certain committed lymphocyte precursors home to the thymus, where they interact with thymic stromal cells and subsequently emerge as mature T cells, expressing new surface markers (differentiation antigens). Surface molecules on lymphocytes can be used to identify functional subpopulations of cells or stages of maturation. Many of the surface glycoproteins of T cells play a key role in cell-to-cell communication and in antigen recognition. Various T cells are known to assist or suppress antibody synthesis by another set of lymphocytes, B cells, or to engage in cytotoxic activity. T cells are the major cell type involved in the defense against virus-infected cells, the rejection of allogeneic cells and tissues, and delayed-type hypersensitivity reactions.

B cells differentiate along a pathway distinct from that for T cells and express their own unique surface molecules. B cells produce immunoglobulin or antibody. Each B cell is genetically precommitted to the production of antibody with a defined structure and, therefore, with a defined specificity. When a B cell proliferates, forming a clone, its daughter cells make precisely the same antibody. As a B cell matures, it expresses a specific cell-surface receptor for antigen. This receptor and the immunoglobulin that the B cell will secrete at maturity have similar molecular structures and identical binding specificities. There is a third group of cells, including dendritic and Langerhans' cells and monocyte-macrophages, that function by presenting antigen to T cells.

Once antibody is produced, it interacts with antigen and a variety of nonspecific factors (i.e., complement, polymorphonuclear leukocytes, and mast cells) that make up the innate defense system. These innate defense system components direct and magnify the neutralizing effects of antibody on a specifically targeted foreign substance. The recruitment of various nonspecific factors in the efferent phase of the immune response greatly amplifies the activity of antibody and is crucial for the development of effective bodily defenses. An example of this interaction between antibody and a nonspecific defense system is the activation of the complement pathway.

The complement system is characterized by an array of peptides, proenzymes, enzymes, receptors, and biologically active effector molecules. This system, which is made up of a minimum of 20 components, interacts through at least two separate pathways (cascades) to achieve a major biological amplification of the action of specific antibody molecules. The fragments of several of these peptides enhance phagocytosis and chemotaxis and alter vascular permeability. The formation of a membrane attack complex (a complex molecule composed of five complement components) on the surface of a eukaryotic cell induces a structural alteration of the plasma membrane, resulting in cytolysis. A similar mechanism has been shown to be effective on some prokaryotic cells. The complement system, operating in conjunction with other components of the innate immune system (i.e., segmented neutrophils and other host systems, including the coagulation and fibrinolytic enzyme systems), can induce an acute inflammatory response in areas of microbial invasion.

Immunologic Specificity

The specificity imparted on the adaptive immune system involves the production and surface display of receptors capable of binding unique epitopes found on foreign cells, microbes, and substances. Structurally different but functionally similar receptors are found on B and T lymphocytes. Each mature B cell has a surface receptor that is encoded by the same genes as the immunoglobulin it is capable of secreting. It is estimated that more than 1 million different immunoglobulin molecules are produced by the B cells of individual mice and humans (and most likely all other mammalian species). When fully differentiated, each B lymphocyte is equipped to produce antibody molecules with a single specificity. This is accomplished by a mechanism of gene rearrangement that occurs during B-cell maturation.

Immunoglobulin molecules (Figure 1-1) are glycoproteins with a basic four-chain structure composed of two large heavy chains, which are covalently linked by disulfide bonds, and two smaller light chains, which are attached to the heavy chains by disulfide bonds. The region where heavy chains are united is called the *hinge region*. Each heavy chain is composed of four subunits or domains that share amino acid homology. Three of these domains have a relatively constant sequence of amino acids, but the fourth, an N-terminal domain, contains a region where the amino acid sequence is variable (between different B cells). Immunoglobulin light chains have two domains: The COOH-terminal domain is constant in amino acid composition;

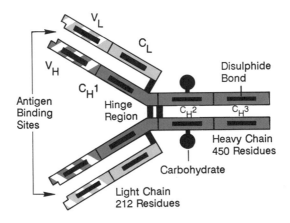

FIGURE 1-1 The basic structure of IgG. The amino-terminal end is characterized by sequence variability (V) in both the heavy (H) and light (L) chains, which are referred to as the V_H and V_L regions, respectively. The rest of the molecule has a relatively constant (C) structure. The constant portion of the light chain is termed the C_L region. The constant portion of the heavy chain is further divided into three structurally discrete regions: C_H1, C_H2, and C_H3. These globular regions, which are stabilized by intrachain disulfide bonds, are referred to as domains. The sites at which the antibody binds antigen are located in the variable domains. The hinge region is a vaguely defined segment of heavy chain between the C_H1 and C_H2 domains. Flexibility in this area permits variation in the distance between the two antigen-binding sites, allowing them to operate independently. Carbohydrate moieties are attached to the C_H2 domains. Source: Roitt et al. (1985). Reprinted courtesy of Professor Roitt, Dr. Brostoff, Dr. Male, and Gower Medical Publishing.

the N-terminal domain is variable. There are five different classes (isotypes) of immunoglobulin (Ig): IgM, IgG, IgA, IgE, and IgD. The IgG class of immunoglobulin in humans, mice, and rats can be broken down into four subclasses, while that in hamsters and guinea pigs has only two known subclasses. Subclasses of IgA are found in humans, mice, and rabbits. Class and subclass distinctions arise from common amino acid sequences found in the heavy chains. In addition, each species has both kappa and lambda light chains, again based on common amino acid sequences. Because immunoglobulin molecules are secreted intact from a B lymphocyte, both heavy chains are identical, and likewise, both light chains are identical. The designation of subclasses and the concentrations of various classes of immunoglobulins vary among mammalian species. Generally, one type of light-chain class predominates in a species. In rats, guinea pigs, mice, and humans the predominant light chain is the kappa chain; in dogs, cats, and horses it is the lambda chain.

Each of the immunoglobulin classes and subclasses has a distinct distribution in the fluid compartments of the body and a distinct function in host defense. The heavy chains of different antibody molecules contain different chemical determinants in the constant and hinge regions of the molecule that activate the complement system, determine the attachment site for antibody on cells, and control the distribution of antibody through the placenta or yolk sac to the embryo.

The genetic mechanism responsible for the tremendous diversity of antibody molecules involves rearrangement of genes for constant (C) regions and variable (V) regions in both the light and heavy chains. Genes that contribute to the peptide sequence of V regions are deployed along the same chromosome that encodes the C regions of each chain. Between the V- and C-region genes are joining (J) and, in some cases, diversity (D) region genes. Genetic sequences encoding for the human kappa-chain C and V regions are on chromosome 2, for the lambda-chain C and V regions are on chromosome 22, and for the heavy-chain C and V regions are on chromosome 14. The genes encoding for the kappa-, lambda-, and heavy-chain C and V regions of mice are located on chromosomes 6, 16, and 12, respectively. The order of gene sequences found on mouse chromosome 12 is $5'$--V--D--J--μ-δ-γ_3-γ_1,γ_{2b},γ_{2a}--ϵ--α--$3'$. The various gene regions found distal to the J region (toward the $3'$ end of the DNA) make up the C-region sequences. Specificity and function of heavy chains are determined by the combination of genes selected from the V, D, J, and C regions. The process involved in coordinating these genetic rearrangements is still not completely understood.

Heavy-chain genes are first to rearrange. This involves joining of a diversity (D) with a J gene, followed by joining of this DJ complex to one of the many V-region genes. Finally the three-member complex VDJ is transcribed as a unit into RNA and, during RNA processing, is spliced adjacent to the C-region genes. The mu (μ) gene for the heavy-chain C region is the first to be chosen. Later in B-cell development, one of the other C-region genes might be chosen through a complex gene switching process.

In both humans and mice, kappa-chain genes are rearranged before lambda-chain genes. Rearrangement of the lambda-chain genes is initiated whenever a given cell fails to make a functional rearrangement at either of its two kappa-chain alleles. After VJ joining occurs, a gene for the C region is selected and light-chain gene expression follows.

Antibody gene rearrangement takes place during the early phases of B-cell differentiation. Once a functional gene rearrangement has occurred within one light- and one heavy-chain cluster, the specificities of both antibody and surface receptor are fixed through the process of allelic exclusion. This mechanism ensures that a B cell will express an immunoglobulin molecule composed of a single light and a single heavy chain. At this point the B cell is ready to bind antigen through its membrane-inserted surface immunoglobulin. Antigen binding is the first in a series of complex steps that lead to the selective proliferation of the B cell and its clonal expansion. Binding of antigen to surface immunoglobulin sets in motion a series of events that include differentiation, as well as proliferation, immunoglobulin secretion, isotype switching, and the generation of immunologic memory. Macrophages, T cells, and lymphokines can influence the differentiation of B cells.

The surface membrane-inserted T-cell antigen receptor (Ti or TCR) is similar to the B-cell antibody. The Ti receptor is a heterodimer, which in most cells is an alpha (α) and a beta (β) chain generated at different stages of T-cell differentiation and inserted through the plasma membrane. A small population of T cells uses a Ti receptor composed of a gamma (γ) and a delta (δ) chain. The Ti receptor is closely associated with a group of T-cell surface molecules collectively called T3. The Ti and T3 glycopeptides appear to work together in the activation of T cells following antigen binding to Ti.

The generation of the Ti receptor uses complex genetic rearrangements of variable-region determinants, which allows for receptor diversity and specificity. The basic mechanism of joining V, D, and C regions is similar to that seen for immunoglobulin heavy chains. The Tiβ-chain gene develops from multiple germline variable-region genes (Vβ) and adjoining sets of diversity (Dβ_1, Dβ_2), joining (Jβ_1, Jβ_2), and constant (Cβ_1, Cβ_2) gene segments. The Tiα-chain gene consists of multiple Vα genes, at least 40 Jα genes, and one constant Cα gene.

The T-cell genes rearrange in an ordered fashion, beginning with the Tiγ- and Tiδ-chain genes and progressing through the Tiα- and Tiβ-chain genes. The function of cells with Ti receptors composed of γ and δ chains remains unclear. The location of the Ti receptor-chain genes includes chromosomes 14 (α and δ), 6 (β), and 13 (γ) in mice. Binding of antigen to a specific Ti receptor is the initial step leading to the generation of clones of fully differentiated T cells. T cells, unlike B cells, do not secrete a soluble form of their antigen receptor.

T cells fall into two or more distinct subpopulations based on functional criteria. T cells, upon maturation, express either helper-inducer or cytotoxic functions and are called Th and Tc cells, respectively. T cells, unlike B cells, are not readily triggered by soluble antigens. Even multimeric antigens or antigens bound to an insoluble matrix are poor stimulators of T cells. In practice, T cells are activated only by antigens associated with the surface of other cells; that is, to be effective, antigen must be "presented" to T cells. The requirements for successful presentation are under intense study, but it is thought that antigen or antigen fragments are effective T-cell activators only when they are bound to certain presenting cell-surface molecules, in particular, to major histocompatibility complex (MHC) antigens. MHC antigens are responsible for evoking tissue rejection responses. Class I MHC antigens are found on all nucleated cells and are composed of a 45-kilodalton (kDa) polypeptide chain and a 12-kDa chain. The latter, a β_2 microglobulin, is highly conserved, while the larger chain is encoded by members of a large gene family with extensive polymorphism. Class II MHC antigens (Ia antigens) are composed of 28-kDa and 32-kDa subunits. Their expression is limited to antigen-presenting cells, including B cells, macrophages, and dendritic cells. In humans, but not in mice, Ia antigens are also expressed on T cells.

Afferent, Central, and Efferent Limbs of the Immune System

The sequence of events that occur from the moment a foreign substance enters the body until it is destroyed or neutralized is highly coordinated and involves both nonspecific and specific immune defenses. To aid in the description of the events confronting the immune system, scientists have designated all events occurring before the activation of T and B lymphocytes as the afferent phase, all activities involving the specific interaction of antigen with lymphocytes as the central phase, and the activity of effector mechanisms in neutralizing or destroying the foreign substance as the efferent phase.

Foreign substances or pathogenic microorganisms sometimes elude the host's physical barriers and enter the body at several different sites. The major portals of entry are the skin (integument) and mucous membranes. It is not surprising, therefore, to discover that specialized defense systems, including immunologic defenses, have been established to deal with intrusions at these sites. Both the skin and mucous membranes are richly endowed with lymphatics that drain to regional lymph nodes. In addition, specialized lymphoid structures, that is, the tonsils, Peyer's patches, and appendix, are found at various sites associated with mucosal surfaces. These are referred to collectively as mucosa-associated lymphoid tissue (MALT). The skin is rich in a variety of specialized cells (mast cells, Langerhans' cells, indeterminate cells, and veiled cells) that participate in immune system function. The last three cell types are known to participate in antigen presentation.

Afferent Limb

Organisms newly introduced into a host are initially confronted by circulating monocytes, macrophages, specific T and B cells, neutrophils, or, in the skin, Langerhans' cells. Some of these cells participate in phagocytosis and degradation (neutrophils, macrophages), and others are nonphagocytic and participate chiefly in antigen presentation (Langerhans' and other dendritic cells). Organisms also enter the lymphatics, either directly or following partial digestion by phagocytes, where they are transported to regional lymph nodes, to MALT, or to both. Foreign substances introduced into the blood confront the reticuloendothelial system, which comprises the spleen, blood vessel endothelium, alveolar macrophages, and Kupffer cells of the liver. The cumulative events that occur from the moment an organism breaches a primary barrier until it is processed, transported, and presented to lymphocytes comprise the afferent limb of the immune system.

Central Limb

During the central phase of the immune response, there is an antigen-induced activation of either specific T-cell or B-cell clones or both. For those antigens that require T- and B-cell interactions to achieve antibody synthesis, T lymphocytes must recognize class II MHC antigens on the surface of antigen-presenting cells. This two-point recognition system, that is, foreign antigen and MHC determinants, allows T and B cells to come into close apposition. Stimulation of Th cells results in the secretion of lymphocyte growth factors interleukin-2 (IL-2), IL-4, and IL-5, which, together with foreign antigen, can induce B-cell proliferation. In this manner the Th cell provides a positive signal that allows for the selective proliferation of a clone of B cells preselected by the specificity for and binding to foreign antigen. Clonal proliferation then results in an expanded population of B cells. Some of these undergo terminal differentiation and produce antibody, in part because of the effects of the lymphokines γ-interferon and IL-6. Others are available for future clonal proliferation. This latter population of B cells provides the basis for immunologic memory and the secondary immune response. Similarly, T cells, following binding to Ia-associated antigen, produce T-cell growth factors (IL-2 and IL-4) and express the appropriate receptors for them.

It has been postulated that another T cell, the Ts cell, can also be specifically activated in a manner similar to that of Th cells. In contrast to Th cells, the Ts cell is thought to secrete a variety of suppressor substances that interfere with B-cell activation and diminish the production of antibody.

The cytotoxic T cell can be activated by mechanisms similar to those described above. This cell enters the circulation in search of a target cell that displays a determinant recognized by its receptor. Its ability to bind and liberate cytocidal substances requires a recognition of antigen in association with class I MHC antigen determinants.

Efferent Limb

As previously mentioned, the binding of antibody to foreign antigen results in a conformational change in the antibody molecule, thus producing an activation site for the first component of complement. The sequential activation of the complement system via the classical pathway generates biologically effective split products that assist in mediating local inflammation. The systemic activation of complement can result in a more injurious activation of the kinin and coagulation pathways, culminating in shock. Complexes of antigen, antibody, and complement are removed quickly from the circulation by cells bearing either Fc receptors, which bind the constant region of IgG or IgM, or complement receptors, which bind the activated products of complement (e.g., CR1).

Antibody-dependent cytotoxic cells (ADCC) are lymphocytes or monocytes that bear the Fc receptor for antibody heavy chains. These cells become armed with specific antibodies and deliver a combined assault known as antibody-dependent cellular cytotoxicity on targeted cells. Similarly, homocytotropic (IgE) antibodies bind mast cells and basophils. Degranulation takes place following the binding of foreign antigen to these cells, resulting in the release of histamine, heparin, and other mediators of inflammation. This mast cell-mediated mechanism is thought to be important in the expulsion of endoparasites from the gastrointestinal tract.

Natural killer (NK) cells apparently recognize and destroy cells (e.g., cancer cells or virus-infected cells) nonspecifically. Some NK cells are large, granular lymphocytes of uncertain lineage. Others are descended from Tc cells and bear T-cell surface differentiation antigens and both Ti and T3.

For all three phases of the immune system to operate normally, a wide variety of other humoral immune modulators must interact properly with cells of both the nonspecific defense system and the immune system. T cells, for instance, are known to have receptors for estrogen. The binding of estrogen to this receptor generates a signal that enhances immunity. NK cells have a dependency on serotonin, which is present in circulating platelets. All cells have receptors for insulin, growth hormone, and thyroxine, and the involvement of these hormones is necessary for properly functioning host defenses. Lymphocytes have receptors for glucocorticoids, which are immunosuppressive. In addition, lymphocytes can have receptors for β-endorphin, prolactin, histamine, calcitonin, and β-adrenergic substances. Receptors for 1,25-dihydroxyvitamin D_3 are not present on resting lymphocytes but are synthesized during lymphocyte activation. Although a precise role for many of these hormones has not been defined, each potentially has an immunomodulating activity.

Immunodeficiency

Compromises in host defenses involve the defective function of either specific immunity or the nonspecific defense mechanisms. The defects in any of the three functional phases of the immune system (afferent, central, or efferent) can compromise immune-mediated defenses. Defects in the immunologic defenses can be either primary or secondary. Primary defects are usually genetically determined, and secondary deficiencies result from developmental abnormalities, environmental factors, or as consequences of microbial action. Both primary and secondary immunodeficiencies compromise greatly the capacity of mammalian hosts to live successfully in a microbe-laden environment.

Primary genetic deficiencies of immunity are rather infrequent as individual abnormalities; however, they are not rare when considered in the aggregate. Indeed, they make up an important component of the inborn errors of metabolism. Secondary immunodeficiencies of humans and animals, occurring as the consequence of malnutrition, cancer, environmental or pharmacologic intoxication, metabolic disorders, pathologic processes, or aging, are among the most frequent underlying causes of serious life-threatening diseases.

The immunodeficiencies of rodents and humans can be visualized in the context of genetic deficiencies in any of several functional components of the immunologic defenses. For purposes of classifying the disorders of immune function, one attempts to define the major immunologic cell systems that are primarily involved. The integrity of primary tissues responsible for lymphocyte development (thymus and bone marrow), the precise lymphoid cell population or subpopulation absent or perturbed in development, the nonspecific cellular defense involved in an abnormality, the molecular basis of the defect, and, finally, the precise genetic basis of the compromised immunologic function are all important in the final classification. Some of the most impressive insights into immunologic arrangements and function have been discovered by studying immunocompromised rodents.

EFFECT OF ENVIRONMENTAL FACTORS ON IMMUNE FUNCTION

Investigators interested in probing the immune system must be aware that certain environmental factors, both infectious and noninfectious, can lead to a transient immune suppression or stimulation. Such factors complicate research results, regardless of whether the animal is normal or immunodeficient, and should be avoided.

Noninfectious Agents

Various agents have been associated with changes in the function of the immune system in rodents, including diet, stress, and drugs. Dietary contaminants such as lead and cadmium increase the susceptibility of rodents to infectious diseases (Hemphill et al., 1971; Cook et al., 1975). Cadmium, in particular, has been associated with suppression of interferon production (Blakley et al., 1980) and abnormalities in macrophage function (Loose et al., 1978a). Abnormalities in both lymphocyte and macrophage activity have been reported in rodents drinking hyperchlorinated acidified water (Fidler, 1977; Hermann et al., 1982).

Various factors that ultimately cause stress in rodents can impair immunologic responsiveness and increased susceptibility to infectious agents. Landi et al. (1982) have demonstrated marked suppression of delayed-type hypersensitivity and specific immunoglobulin production and elevation in corticosterone production that persist in mice for 48 hours following arrival after shipment. Furthermore, the stress associated with shipping is known to cause perturbances in a wide range of hematologic and biochemical parameters in rats (Bean-Knudsen and Wagner, 1987). Overcrowding of mice or rats has also been associated with abnormalities in immunity (Baker et al., 1979b; Riley, 1981). Increased susceptibility to certain infectious agents has been related to alterations in temperature (Baetjer, 1968) and to changes in light–dark cycles, with a subsequent perturbance in their circadian rhythm (Pakes et al., 1984).

The role of various drugs, particularly antibiotics, on immunity has been reviewed (Hauser and Remington, 1982), as has the effect of drugs on animal physiology (Hsu, 1976; Pakes et al., 1984). Tetracyclines have been shown to depress delayed-type hypersensitivity reactions and lymphocyte transformation in mice (Thong and Ferrante, 1980). Various anesthetic agents have been associated with depressed immunity and increased tumor cell metastasis in rodents (Shapiro et al., 1981). Halothane is known to decrease the response of rat lymphocytes to nonspecific mitogens (Bruce, 1972) and to depress chemotaxis and phagocytosis (Cullen, 1974).

Infectious Agents

A wide range of pathogenic agents has been associated with changes in immunity that will complicate research with immunodeficient rodents. These have been reviewed extensively (Wagner and Manning, 1976; Baker et al., 1979a; Hsu et al., 1980; Foster et al., 1982; Fox et al., 1984). Profound changes in immunity have been seen in laboratory rodents infected subclinically with Sendai virus (Garlinghouse and Van Hoosier, 1978; Kay, 1978), mouse hepatitis virus (MHV) (Callisher and Rowe, 1966), lactic dehydrogenase virus (LDV) (Riley et al., 1978), lymphocytic choriomeningitis virus (LCMV) (Oldstone and Dixon, 1971), murine cytomegalovirus (Doody et al., 1986), minute virus of mice (MVM) (Bonnard et al., 1976), and *Hemobartonella muris* and *Eperythrozoon coccocides* (Baker et al., 1971). Many infectious agents (e.g., Sendai virus, MHV, reovirus 3, LDV, LCMV, and MVM) are found as contaminants in transplantable tumors and tumor cell lines (Stanley, 1965; Collins and Parker, 1972; Biggar et al., 1976), which are commonly inoculated into immunologically compromised hosts.

Murine cytomegalovirus, *E. coccocides*, *H. muris*, LDV, reovirus 3, Sendai virus, and others, although themselves clinically silent, predispose the host to more severe infections by other agents (Klein et al., 1969; Richter, 1970; Baker et al., 1971; Stanley and Joske, 1975; Hamilton et al., 1976;

Riley and Spackman, 1977). Some rodent pathogens cause clinically silent infections that can be activated by concurrent infection with another agent or by an experimental procedure (Callisher and Rowe, 1966; Baker et al., 1971). Strong synergistic effects have been observed between Sendai virus and other viral, bacterial, and mycoplasma pathogens (Saito et al., 1978a; Jakab, 1981). Likewise, K virus is known to potentiate the effects of MHV in weanling mice (Tisdale, 1963). At least one agent, LCMV, is known to be a zoonotic virus responsible for life-threatening disease in humans (Biggar et al., 1976).

Some infectious agents of rodents are known to derepress oncogenic virus genes (Riley and Spackman, 1977; Zinkernagel et al., 1977); others markedly promote or suppress the growth of experimental tumors (Nadel and Haas, 1956; Hotchin, 1962; Molomut and Padnos, 1965; Riley, 1966; Bonnard et al., 1976; Riley et al., 1978; Peck et al., 1983). In addition, the response to chemotherapeutic drugs can be altered by simultaneous infection with LDV (Riley et al., 1974, 1978).

Infection with oncogenic retroviruses (oncornaviruses) is associated with a significant immunosuppressive action. For example, in mice, infection with Gross virus produces a persistent, progressive, and profound inhibition of antibody production and a depression of cell-mediated immunities, including skin allograft rejection (Dent, 1972). Similarly, murine Friend leukemia virus complex, Moloney virus, Raucher virus, and mammary tumor viruses (Siegel and Morton, 1966; Bennett and Steeves, 1970); feline leukemia virus (FeLV) (Cockerell, 1976); and human T-cell lymphotropic virus type I (HTLV-I) (Popovic et al., 1984) exert immunosuppressive influences.

Recent evidence indicates that at least some of the viral-associated immunosuppressive influences (i.e., suppression of antibody production, T-cell-mediated immunities, interferon production, and bactericidal functions of phagocytic macrophage cells) can be attributed to a transmembrane portion of the envelope component called P-15E, which is present in each of these oncornaviruses (Mathes et al., 1978, 1979; Snyderman and Cianciolo, 1984). Furthermore, a highly conserved small peptide, CKS17, located within the P-15E sequence has been identified and synthesized. This small peptide, when linked to human albumin, exerts a powerful immunosuppressive action in vitro. The immunosuppressive action of CKS17 might account, at least in part, for the immunosuppressive actions of the oncornaviruses (McChesney and Oldstone, 1987).

Lentiviruses (cytopathic retroviruses that do not cause cancer) are potent immunosuppressive viruses in humans (Lane et al., 1985), monkeys (Stromberg et al., 1984), and cats (Pedersen et al., 1987). However, they have not been described in rodents.

Certain viruses (e.g., LCMV) alter the course of spontaneous autoimmune disease (Tonietti et al., 1970; Oldstone, 1988), and chronic infections by reovirus 3, encephalomyocarditis virus, coxsackievirus B4, and LCMV are associated with the development of diabetes mellitus (Yoon et al., 1978, 1980; Onodera et al., 1981; Oldstone et al., 1984). Infection with the murine pinworm *Syphacia obvelata* decreases the incidence of adjuvant arthritis in rats (Pearson and Taylor, 1975). Infection of mice with reovirus 1 leads to an autoimmune polyendocrine disease, characterized by autoantibodies that react with glucagon, insulin, and growth hormone (Haspel et al., 1983). There is mounting evidence that in humans, as well as in laboratory rodents, some autoantibodies characteristic of autoimmune disease are anti-idiotype, anti-viral antibodies (Plotz, 1983). In addition, certain autologous antigens, for example, thyroid-stimulating hormone (TSH) receptor, appear to be antigenically identical to haptenic groups on various infectious agents (Weiss et al., 1983; Stefansson et al., 1985). For these reasons investigators must maintain laboratory animal models free from infection by pathogens.

GENERAL CONSIDERATIONS FOR MAINTAINING IMMUNODEFICIENT RODENTS

An investigator desiring to work with immunodeficient rodents should first be familiar with standard husbandry practices in conventional, specific-pathogen-free (SPF), defined-flora, and germfree environments (Simmons et al., 1968; ILAR, 1970; Bleby, 1976; Dinsley, 1976; Trexler, 1976, 1983; Canadian Council on Animal Care, 1980; Sasaki et al., 1981; Otis and Foster, 1983; NRC, 1985). The overall purpose is to prevent exposure to agents that are detrimental to the host or that can alter experimental results. The protective barrier must be consistent with the nature of the facility and the experimental objectives. Under environmental conditions in which the profile of existing microorganisms is well defined or pathogenic agents are absent, minimal protection is necessary to maintain most immunodeficient rodents. Under environmental conditions in which the microbiological status of the environment is unknown or pathogenic agents are present, strict isolation should be used.

Most immunodeficient rodents do not require barrier maintenance; however, all will benefit from being housed in a pathogen-free environment. The general conditions outlined in the *Guide for the Care and Use of Laboratory Animals* (NRC, 1985) should be used for the maintenance of these animals. Certain immunodeficient rodents [i.e., nude (*nu*) mice, rats, and hamsters; mice with severe combined immune deficiency (*scid*); and any multiple mutant strains homozygous for the *nu* gene] are susceptible to a broad array of indigenous agents. For these mutants strict isolation is obligatory. The care of these animals is detailed in Chapter 4.

While exposure to infectious agents is a major concern for anyone dealing with immunodeficient rodents, some of the strains discussed in this volume have additional physiologic abnormalities. Many of these nonimmunologic deficits are associated with chronic debilitation and reduced life expectancy, and it can be assumed that some animals will experience pain and distress. This document addresses the nonimmunologic deficits separately and provides guidelines for the proper care of these animals. It is an additional obligation of investigators who use these animals to monitor their well-being regularly; to minimize their suffering whenever possible; and, when necessary, to perform euthanasia in conformity with the recommendations of the American Veterinary Medical Association Panel on Euthanasia (1986).

MUTATIONS

The determination of cause-and-effect relationships between normal and abnormal physiologic processes is often complex, and specific experimental techniques are required to study them. One effective technique is to study systems in which normal function has been altered by mutation, a spontaneous heritable change that results in some measurable change in the structure or function of the organism. The optimal use of the mutant models described in this guide requires an understanding of basic genetic principles. The General Reading section at the end of this chapter should be consulted for relevant references.

The expression of a recessive mutation requires the presence of a mutated allele (gene) on both the maternal and paternal chromosomes. Codominant or semidominant mutations require only one gene for expression of the trait; however, both parental phenotypes are expressed in the heterozygote. Dominant mutations require only one mutant gene for full expression. Knowledge of the type of mutation is essential for setting up proper mating systems. By convention, symbols for recessive genes are written in all lowercase letters, and those for codominant and dominant genes are written with the first letter capitalized. When used in scientific publications, the symbols must be underlined or italicized. Information on obtaining the rules of standardized nomenclature for rodents is given later in this chapter.

The advantages of mutations are that their effects can be precisely determined and that they are, for the most part, inherited in a simple and predictable pattern. In addition, the phenotypic expression of certain mutant loci is similar in divergent species. For this reason many spontaneously occurring rodent mutations result in models for human diseases, for example, beige mice and humans with Chédiak-Higashi syndrome. Their disadvantage is that establishing their full potential requires time, resources, and work (e.g., to transfer a mutation from one genetic background to another).

Probably the most critical decision to be made in working with a specific mutant is choosing the genetic background(s) on which it should be maintained. The ideal choice is an inbred strain. A mutant gene can be established on an inbred strain by the use of appropriate mating systems. The result is defined as a congenic strain, that is, a strain that is genetically identical to its partner inbred strain, with the exception of the locus in question and its closely linked genes. A great deal can be learned about the function of a specific gene by comparing the congenic strain with its inbred partner. This technique also allows the investigator to determine whether the heterozygote differs from either homozygote (mutant or wild type).

The next consideration is the choice of the strain. The expression of a mutation can be modified by the background genetics of the strain. Thus, if a mutant appears to be particularly valuable, transfer of the gene to two or more strains should be considered. This provides an additional dimension, as it becomes possible to study the effect of a particular mutant on two different genetic backgrounds under the same experimental conditions. The number of strains chosen is limited only by the available resources and the imagination of the investigator.

Although individual mutants provide a powerful tool for the study of specific events, genes rarely function independently. Therefore, the next step in the evolution of an experimental design is to combine two or more genes affecting the same system on the same inbred background. If two genes are combined, the result is an experimental design that allows four comparisons; if three genes are combined, eight comparisons result. Studies using these combinations contribute to our understanding of how genes interact.

GENE MARKERS AND CHROMOSOME MAPS

Whenever known, the chromosomal location of each mouse mutation discussed in this report is provided. The chromosome map (Figure 1-2 [pp. 18–31], see caption on p. 17) of the mouse can be used to find linked genes that can serve as markers to identify affected animals prior to gene expression, even while they are still in the embryonic stage, or to distinguish heterozygous carriers. Linkage can also be useful in studies of gene regulation and interspecies chromosome homologies. Genes are being added to the mouse chromosome map at a rapid rate, and if linked genes are being sought, up-to-date knowledge should be obtained. Information on mouse linkage is available from Drs. M. T. Davisson, T. H. Roderick, A. L. Hillyard, and D. P. Doolittle of the Jackson Laboratory, Bar Harbor, Maine 04609 (207-288-3371).

Unlike the mouse, most of the rat mutations discussed in this report have not yet been mapped. Nonetheless, a linkage map of the rat has been included (Figure 1-3) to aid users of known rat mutants, as well as those who discover new genetic variants, in elucidating linkage associations. Additional information on rat linkage is available from Dr. H. J. Hedrich, Central Institute for Laboratory Animal Breeding, Hannover, Federal Republic of Germany.

NOMENCLATURE AND SOURCES OF IMMUNODEFICIENT RODENTS

To avoid ambiguity in identifying genetically defined strains of rodents, internationally accepted rules of standardized nomenclature have been established. The importance of this standardized nomenclature cannot be overemphasized. It defines the basic genetic characteristics of the animal and enables investigators in separate laboratories to compare their experimental results. Listing only the strain name of an animal in a research paper is not sufficient, because the genetics of different substrains of an inbred strain (e.g., the C3H strain of mice) can vary dramatically. As an example, C3H/HeJ mice have a defect in lipopolysaccharide response that is not found in the C3H/HeN or any other C3H substrain. Even listing the strain and source is not sufficient, because a single source might hold several different substrains. Similarly, describing a mutation only by the gene name is inadequate because the background on which a mutant gene is carried can affect gene expression. For instance, the effects of the mutations obese (*ob*) and diabetes (*db*) differ depending on the genetic background. C57BL/Ks mice carrying either of these mutations show decreased plasma insulin and degeneration of the islets of Langerhans. C57BL/6J mice carrying either of the same mutations show high plasma insulin levels and hypertrophy of the islets of Langerhans.

(Text continues on page 34.)

FIGURE 1-2 (pages 18–31) Chromosome map of the mouse. Solid vertical bars represent the chromosomes. They are drawn to their proportional lengths based on an estimated total haploid length of 1,600 cM. Centromeres are represented by knobs. Nucleolus organizers are symbolized by **NO**, except on chromosome 12, where a ribosomal RNA gene has been mapped using a DNA polymorphism and is symbolized by **Rnr12**. Gene symbols are given to the right of the chromosome bars; recombination percentages or cM are given to the left. Distances are given as cM from the centromere. The resolution of the current map is 1 cM. The distances between centromeres and proximal markers in most chromosomes have been determined using Robertsonian chromosomes and might be underestimated. Genes listed at the bottoms of chromosomes have been assigned to those chromosomes by parasexual methods or are known only to be linked to those chromosomes. The map is compiled from female and male linkage data and recombinant inbred strain data. When estimated distances for the same interval differ significantly, a weighted average has been used. Anchor loci, whose positions are well known from three-point crosses and extensive data, are shown by long lines extending through the chromosome bars and boldface symbols. Loci whose positions are known with less assurance are indicated by shorter lines. When a locus is mapped with respect to only one other locus, the line is drawn to, but not through, the chromosome, and the symbol of the locating locus is added in parentheses. When a locus is known to be near another locus, but recombination values are not known, the new locus is placed next to the linked locus but no line is drawn to the chromosome. When more than one locus maps to the same position, the loci are listed on the same line, and if one is an anchor locus, its symbol is given first. An upward caret (/\) means that the locus or loci following it belong in the line above. The symbols => or <= indicate order; the arrow points to the more distal locus. Shaded circles indicate genes responsible for the immunodeficiencies discussed in this report. Shaded rectangles indicate other genes mentioned in the text. Map compiled by M. T. Davisson, T. H. Roderick, A. L. Hillyard, and D. D. Doolittle, The Jackson Laboratory, Bar Harbor, Maine (1989).

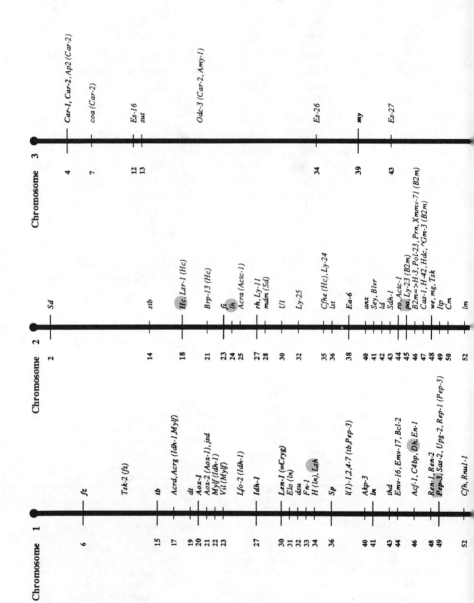

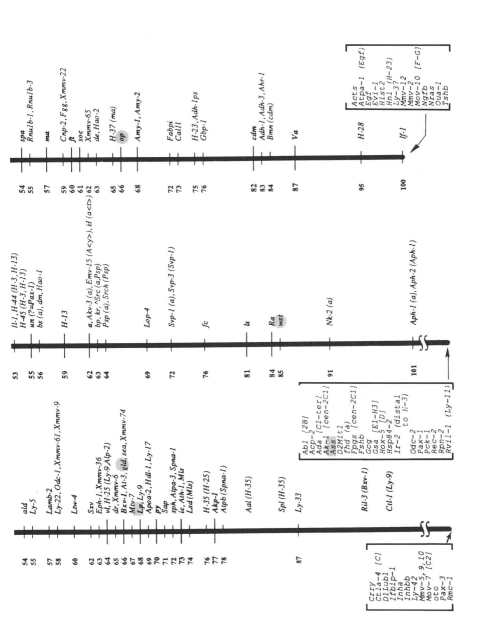

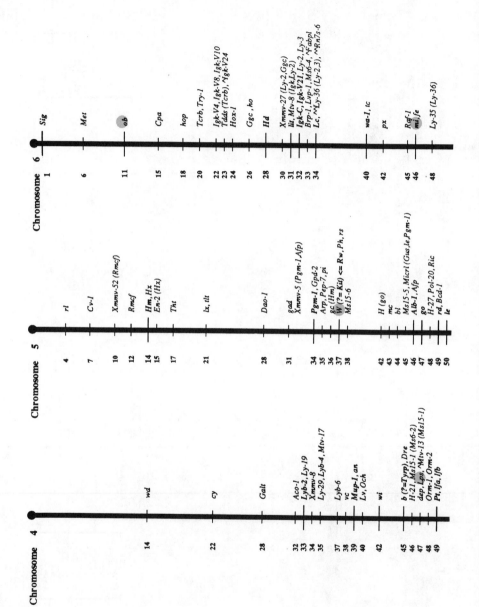

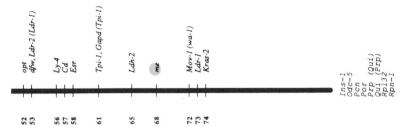
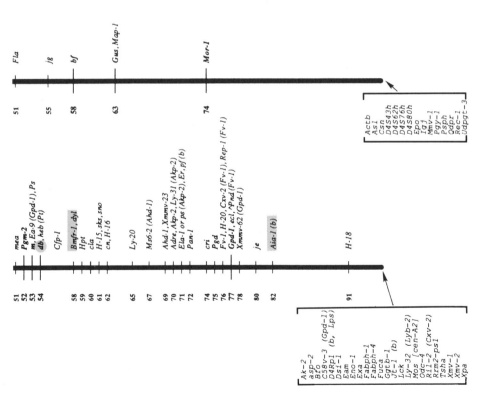

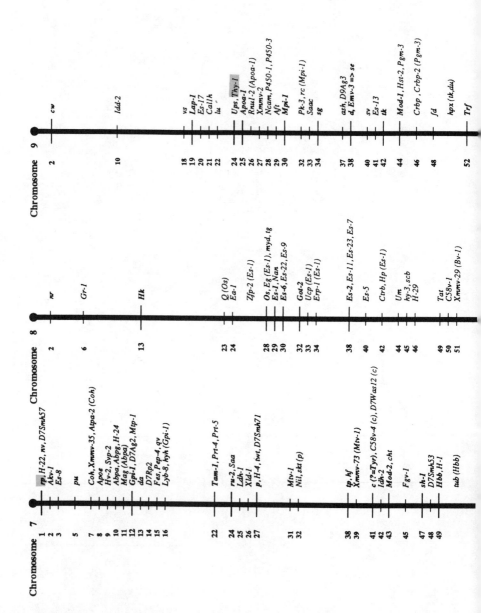

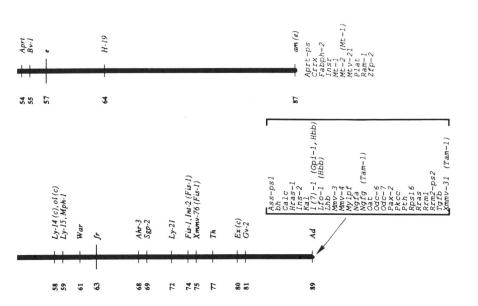

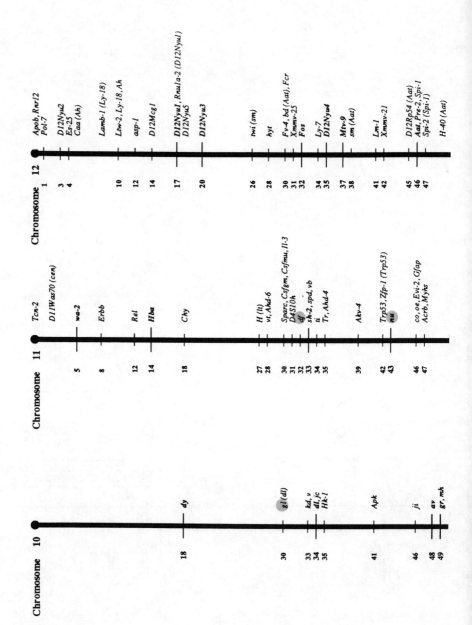

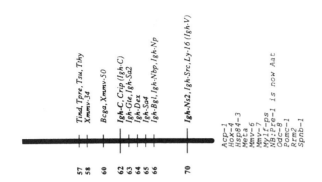
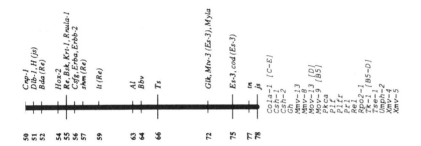
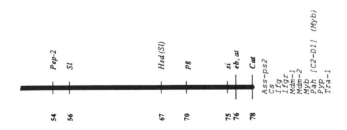

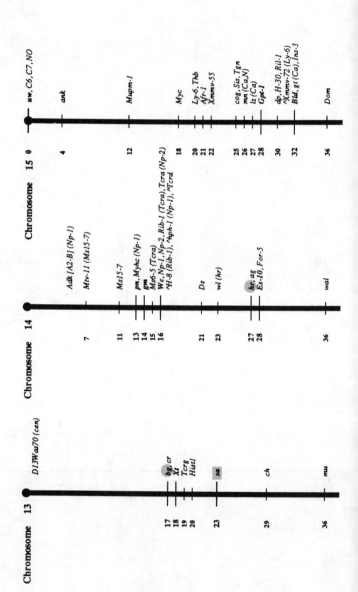

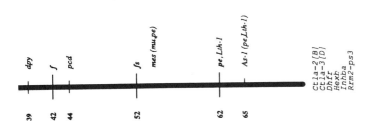

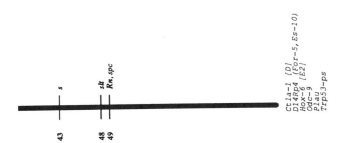

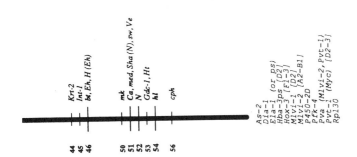

28

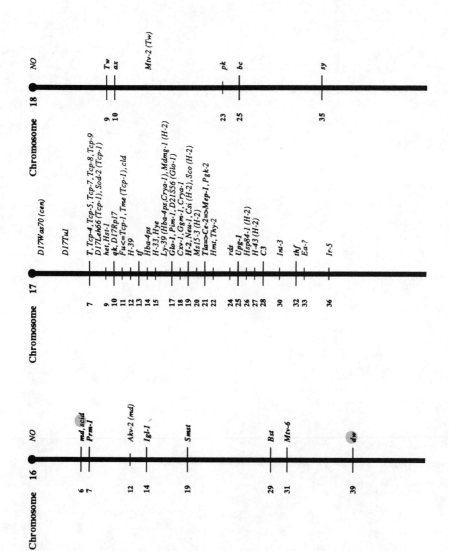

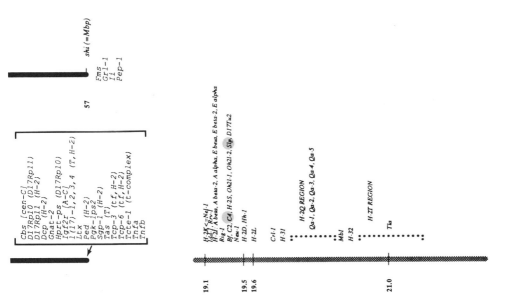
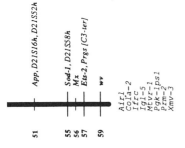

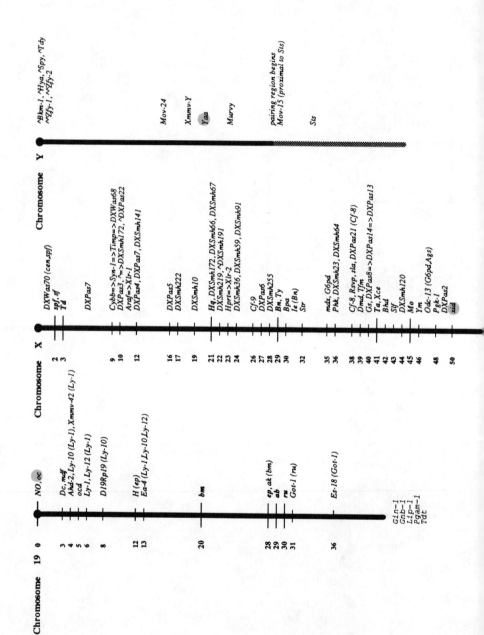

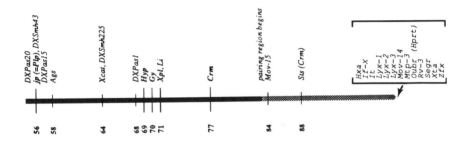

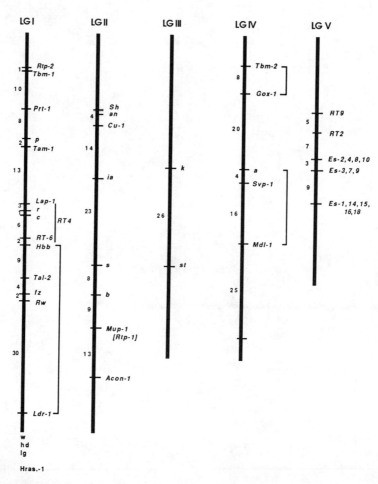

Chromosome 1

FIGURE 1-3 Linkage map of the rat, which is based on literature and recent unpublished information. Except where indicated, the linkages shown have not yet been assigned to chromosomes. An enlargement of *RT1* is given below linkage group IX ("★"). Linkage data for *Es-6, Ir-JHM, RT3*, and *Tbm-2* have been established by recombinant inbred strain analysis. Brackets indicate that the position of one of the two genes with respect to other genes within this linkage group has not been verified. The assignment of linkage groups I, VIII, and IX to chromosomes 1, 6, and 14, respectively, is based on Levan et al. (1986). Map compiled by H. J. Hedrich, Central Institute for Laboratory Animal Breeding, Hannover, Federal Republic of Germany (1988).

INTRODUCTION 33

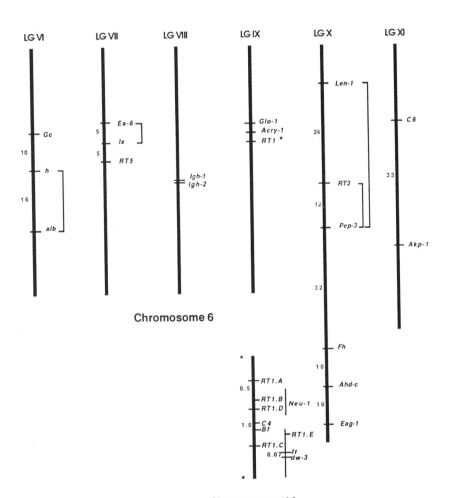

Chromosome 6

Chromosome 14

In general, the rules of standardized nomenclature followed for all the commonly used laboratory rodents are those prepared by the International Committee on Standardized Genetic Nomenclature for Mice. The most recent version of these rules will be available shortly for both mice (Lyon, in press) and rats (Greenhouse et al., in press). Updates of the rules are printed periodically in *Mouse News Letter*, which is published by Oxford University Press, Walton Street, Oxford OX2 6DP, England, and in *Rat News Letter*, which is edited and printed by Dr. Donald V. Cramer, 712 Scaife Hall, University of Pittsburgh School of Medicine, Pittsburgh, PA 15261.

In designating hybrids (the first-generation offspring of a cross between two inbred strains), it is customary to list the female parent first. Thus, the first-generation offspring of a cross between a C57BL/6J female and a C3H/HeJ male is written C57BL/6J × C3H/HeJ F_1 or, if necessary for clarity, (C57BL/6J × C3H/HeJ)F_1. The filial generation (F) number is written as a subscript to distinguish it from the strain and substrain designations. Once the hybrid has been fully described in this manner, an abbreviation may be used thereafter to save time and space. If an abbreviation is used, it should be appended to the initial designation, for example, C57BL/6J × C3H/HeJ F_1 (hereafter called B6C3F_1). A list of abbreviations for inbred strains can be found in Staats (1981).

No sources are given for the immunodeficient rodents discussed in this report because of the rapidly changing nature of commercial and research animal colony holdings. To locate these models, contact the Animal Models and Genetic Stocks Information Program of the Institute of Laboratory Animal Resources, National Research Council, 2101 Constitution Avenue, NW, Washington, DC 20418 (202-334-2590).

GENERAL READING

The following general references provide information on the topics discussed in this report:

Altman, P. L., and D. D. Katz, eds. 1979. Inbred and Genetically Defined Strains of Laboratory Animals. Part 1. Mouse and Rat. Bethesda, Md.: Federation of American Societies for Experimental Biology. 418 pp.

Benjamin, E., and S. Leskowitz. 1988. Immunology: A Short Course. New York: A. R. Liss. 328 pp.

Feldman, D. B., and J. C. Seely. 1988. Necropsy Guide: Rodents and the Rabbit. Boca Raton, Fla.: CRC Press. 167 pp.

Fogh, J., and B. C. Giovanella, eds. 1978, 1982. The Nude Mouse in Experimental and Clinical Research. New York: Academic Press. 1978, vol. 1, 502 pp.; 1982, vol. 2, 587 pp.

Gershwin, M. E., and E. L. Cooper, eds. 1978. Animal Models of Comparative and Developmental Aspects of Immunity and Disease. New York: Pergamon. 396 pp.

Gershwin, M. E., and B. Merchant, eds. 1981. Immunologic Defects in Laboratory Animals. New York: Plenum. Vol. 1, 360 pp.; vol. 2, 382 pp.

Gill, T. J., III, H. W. Kunz, D. N. Misra, and A. L. Cortese-Hassett. 1987. The major histocompatibility complex of the rat. Transplantation 43:773–785.

Green, E. L., ed. 1966. Biology of the Laboratory Mouse, 2nd ed. New York: McGraw-Hill. 706 pp.

Green, E. L. 1981. Genetics and Probability in Animal Breeding Experiments. New York: Oxford University Press. 271 pp.

Green, E. L., and D. P. Doolittle. 1963. Systems of mating used in mammalian genetics. Pp. 3–41 in Methodology in Mammalian Genetics, W. J. Burdett, ed. San Francisco: Holden-Day.

Hedrich, H. J., ed. In press. Genetic Monitoring of Inbred Strains of Rats. A Manual on Colony Management, Basic Monitoring Techniques, and Genetic Variants of the Laboratory Rat. Stuttgart: Gustav Fischer Verlag.

Klein, J. 1982. Immunology: The Science of Self-Nonself Discrimination. New York: John Wiley & Sons. 687 pp.

Klein, J. 1986. Natural History of the Major Histocompatibility Complex. New York: Wiley Interscience. 775 pp.

Lyon, M. F. 1963. Genetics of the mouse. Pp. 199–234 in Animals for Research. Principles of Breeding and Management, W. Lane-Petter, ed. London: Academic Press.

Lyon, M. F., and A. G. Searle, eds. In press. Genetic Variants and Strains of the Laboratory Mouse, 2nd ed. Oxford: Oxford University Press.

Paul, W. E. 1984. Fundamental Immunology. New York: Raven Press. 809 pp.

Reed, N. D. 1982. Proceedings of the Third International Workshop on Nude Mice. Vol. 1: Invited Lectures, Infection, Immunology; 330 pp. Vol. 2: Oncology; 690 pp. New York: Gustav Fischer.

Roitt, I., J. Brostoff, and D. Male. 1985. Immunology. London: C. V. Mosby, Gower Medical. 300 pp.

Salzman, L. A. 1986. Animal Models of Retrovirus Infection and Their Relationship to AIDS. Orlando, Fla.: Academic Press. 470 pp.

Shultz, L. D., and C. L. Sidman. 1987. Genetically determined murine models of immunodeficiency. Annu. Rev. Immunol. 5:367–403.

Sordat, B., ed. 1984. Immune-Deficient Animals. Basel: S. Karger. 445 pp.

Theofilopoulos, A. N., and F. J. Dixon. 1985. Murine models of systemic lupus erythematosus. Adv. Immunol. 37:269–390.

2

Hereditary Immunodeficiencies

This chapter describes the genetics, pathophysiology, husbandry, and reproduction of 64 inbred, hybrid, and mutant strains of rodents with hereditary immunodeficiencies. It is not an exhaustive review, but rather, it summarizes current knowledge and provides key references as a starting point for the reader. Most of the animals described do not need extraordinary methods of isolation and maintenance, but do require procedures that prevent the introduction of rodent pathogens. The procedures necessary for the care of those animals susceptible to ubiquitous organisms are detailed in Chapter 4. Strains to be discussed are listed in Table 2-1.

MICE WITH SINGLE MUTATIONS

***Bcg^r, Bcg^s* (Resistance and Susceptibility to *Mycobacterium bovis*); *Ity^r, Ity^s* (Resistance and Susceptibility to *Salmonella typhimurium*); *Lsh^r, Lsh^s* (Resistance and Susceptibility to *Leishmania donovani*)**

Genetics

Resistance and susceptibility alleles of *Bcg*, *Ity*, and *Lsh* define relative resistance to in vivo infection with *Mycobacterium bovis* (Gros et al., 1981) or *M. lepraemurium* (Skamene et al., 1984; Brown and Glynn, 1987), *Salmonella typhimurium* (Plant and Glynn, 1979), and *Leishmania donovani* (Bradley et al., 1979), respectively. Phenotyping of several inbred, recom-

TABLE 2-1 List of Immunodeficient Rodents

Mutation or Strain	Page No.	Dysfunctions Other than Immunodeficiency		Care	
		Auto-immunity	Non-immune	Special Breeding Techniques	Special Husbandry Procedures
Mice with single mutations					
Bcg^r; Bcg^s (resistance or susceptibility to *Mycobacterium bovis*)	36	−	−	−	−
bg (beige)	40	−	+	−	−
db (diabetes)	43	+	+	+	+
df (Ames dwarf)	46	−	+	+	+
Dh (dominant hemimelia)	47	−	+	+	−
dw (dwarf)	49	−	+	+	+
gl (grey-lethal)	51		+	+	−
gld (generalized lymphoproliferative disease)	52	+	−	−	−
Hc^0 (hemolytic complement)	54	−	−	−	−
hr (hairless); hr^{rh} (rhino)	55	−	+	+	−
Ity^r; Ity^s (resistance or susceptibility to *Salmonella typhimurium*)	36	−	−	−	−
lh (lethargic)	58	−	+	+	−
lpr (lymphoproliferation)	59	+	−	+	−
Lps^d (lipopolyaccharide response, defective)	62	−	−	−	−
Lsh^r; Lsh^s (resistance or susceptibility to *Leishmania donovani*)	36	−	−	−	−
me (motheaten); me^v (viable motheaten)	64	+	+	+	−
mi (microphthalmia)	67	−	+	+	+
nu (nude); nu^{str} (streaker)	69	+	+	+	+
ob (obese)	72	+	+	+	−
oc (osteosclerotic)	73	−	+	+	+
op (osteopetrosis)	75	−	+	+	+
scid (severe combined immunodeficiency)	77	−	−	−	+
Tol-1 (tolerance to bovine gamma-globulin)	80	−	−	−	−

Continued

TABLE 2-1 (continued)

Mutation or Strain	Page No.	Dysfunctions Other than Immunodeficiency		Care	
		Auto-immunity	Non-immune	Special Breeding Techniques	Special Husbandry Procedures
Mice with single mutations (continued)					
vit (vitilago)	80	−	+	−	−
W (dominant spotting); Wv (viable dominant spotting)	82	−	+	+	−
wst (wasted)	84	−	+	+	−
xid (X-linked immune deficiency)	85	−	−	−	−
Yaa (Y-linked autoimmune accelerator)	87	+	+	−	−
S-region-linked genes controlling murine C4	89	−	−	−	−
Mice with multiple mutations					
gld xid	90	±	−	−	−
lpr nu	90	±	+	+	+
lpr xid	91	+	−	−	−
lpr Yaa	91	+	−	±	−
nu bg	91		+	+	+
nu xid	92	−	+	+	+
nu bg xid	92		+	+	+
nu Dh	93	+	+		+
Yaa bg	93	±	+	−	−
Yaa xid	94	±	−	−	−
Inbred mice					
BSVR, BSVS	94	−	−	−	−
BXSB/Mp (females)	96	+	−	−	−
DBA/2Ha	98	−	−	−	−
MRL/Mp	98	+	−	−	−
NOD	100	+	+	−	+
NON	103	±	+	−	−
NZB	104	+	−	−	−
NZB × NZW F1 (BWF1)	108	+	−	−	−
NZB × SWR F1 or SWR × NZB F1, both abbreviated SNF1	112	+	−	−	−
PN	114	+	−	−	−
SAM-P	115	−	+	−	−
SJL/J	117	+	±	−	+
SL/Ni	120	+	−	+	−

TABLE 2-1 (continued)

Mutation or Strain	Page No.	Dysfunctions Other than Immunodeficiency		Care	
		Auto-immunity	Non-immune	Special Breeding Techniques	Special Husbandry Procedures
Outbred mice					
SWAN	122	+	−	−	−
Rat mutants					
ia (incisor absent)	123	−	+	+	+
op (osteopetrosis)	124	−	+	+	+
rnu (Rowett nude); *rnu^N* (*nznu*, New Zealand nude)	125	−	+	+	+
tl (toothless)	128	−	+	+	+
C4 deficiency	129	−			−
Inbred rats					
BB/Wor	130	+	+	−	+
LOU/C	134	+	−	−	−
Guinea pig mutants					
C2 deficiency	134	+	−	−	−
C3 deficiency	135	−	−	−	−
C4 deficiency	136	+	−	−	−
Hamster mutants					
nu	138	−	−	+	+
C6 deficiency	139			−	−

NOTE: A discussion of each positive (+) listing follows in the narrative describing each specific mutant or strain.

binant inbred, and congenic strains indicates that the resistance and susceptibility alleles for all of these unrelated, obligate, intracellular pathogens are identical or closely linked on chromosome 1 (Plant et al., 1982). The BALB/c, C57BL/6, C57BL/10ScSn, and DBA/1 strains carry the susceptibility allele(s); the A/J, C3H/HeJ, C57L, and DBA/2 strains have the resistance allele(s). BALB/c (Potter et al., 1983) and C57BL/10ScSn (Blackwell, 1985) congenic strains carrying the resistance allele(s) have been developed using DBA/2 and C57L, respectively, as donor strains.

Although these loci do not affect the immune response per se, they are unique in affecting macrophage function. There are other loci that affect resistance to *Mycobacterium bovis*, *M. lepraemurium*, *Salmonella typhimurium*, and *Leishmania donovani*; however, these have not yet been mapped (Curtis et al., 1982; Curtis and Turk, 1984).

Pathophysiology

Short-term culture of splenic and hepatic cells indicates that the phenotype is expressed in macrophages, which vary in their ability to control the rate of intracellular replication of pathogens (Lissner et al., 1983). Denis et al. (1988) evaluated the efficiency of mouse splenic macrophages from strains congeneic for the *Bcg* locus and found that macrophages from Bcg^r mice were superior to those from Bcg^s mice in presenting bacteria-derived soluble and particulate antigens.

In addition to imparting resistance to infection to *Salmonella typhimurium* and *Leishmania donovani*, the *Bcg* locus plays a less significant role in the protection of the closely related agents *Salmonella typhi* and *Leishmania major*. In the case of *L. major*, Davies et al. (1988) found that the parasite has developed a sophisticated method for escaping the effects of the *Bcg* gene. It induces increased monocyte infiltration into the site of infection, thereby providing safe targets in which the parasite can survive and multiply.

Husbandry

Special husbandry procedures are not required.

Reproduction

These animals reproduce normally.

bg (Beige)

Genetics

Beige (formerly also called slate) is a recessive mutation located on chromosome 13 that arose independently several times. The symbol *bg* was assigned to a mutation, probably radiation induced, found at Oak Ridge National Laboratory, Oak Ridge, Tenn. (Kelly, 1957). A spontaneous mutation called slate (*slt*) was discovered at Brown University in 1955 (Chase, 1959) and was first described in 1963 (Pierro and Chase, 1963). In 1965 *slt* was recognized to be allelic with *bg*, and Chase (1965) recommended that the symbol *slt* be dropped if the mutant proved to be identical to *bg*. Another mutant allele, bg^J, occurred spontaneously in the C57BL/6J strain at the Jackson Laboratory, Bar Harbor, Maine (Lane, 1962). Beige-2J (bg^{2J}) arose spontaneously in strain C3H/HeJ at the Jackson Laboratory in 1972.

HEREDITARY IMMUNODEFICIENCIES   41

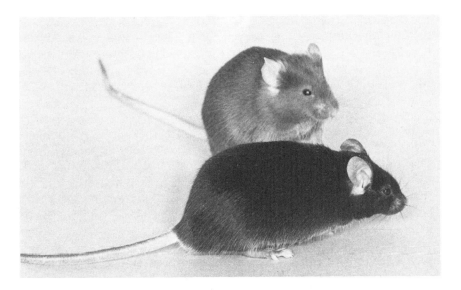

A C57BL/6J-*bg/bg* mouse (rear) and a C57BL/6J control (front). Mice homozygous for the beige mutation show a dilution in pigmentation. Photograph courtesy of the Jackson Laboratory, Bar Harbor, Maine.

Pathophysiology

Homozygous beige mice have a light coat color and reduced ear and tail pigmentation. Eye color is light at birth, changing to a color that varies from ruby to almost black in adults. The phenotypic manifestations in *bg/bg* mice closely resemble the Chédiak-Higashi syndrome in humans, the Aleutian trait in mink, and a syndrome of partially albino Hereford cows (Lutzner et al., 1967).

Beige homozygotes have been reported to have cytotoxic T-cell and macrophage defects (Mahoney et al., 1980; Saxena et al., 1982; Halle-Pannenko and Bruley-Rosset, 1985). They have abnormally large lysosomal granules in a wide variety of cells, including leukocytes of bone marrow and peripheral blood, thyroid follicular cells, type II pneumocytes, mast cells, pyramidal cells and pericytes of the cerebral cortex, Purkinje cells of the cerebellum, spinal cord neurons, islet and acinar cells of the pancreas, liver parenchymal cells, and proximal tubule cells of the kidneys (Oliver and Essner, 1973; Chi et al., 1978; Prueitt et al., 1978). The defective granules are thought to contribute to lowered chemotaxis and a general motility defect (Gallin et al., 1974).

Peritoneal exudate leukocytes of C57BL/6J-*bg/bg* mice contain very low or undetectable levels of neutral protease (Vassalli et al., 1978), and circulating polymorphonuclear leukocytes (PMNs) in *bg* homozygotes have impaired chemotactic and bactericidal activities. Kaplan et al. (1978) showed that the addition of small numbers of normal platelets or serotonin to whole blood of beige homozygotes increased the bactericidal activity of the PMNs. They theorized that normal platelets or serotonin enhanced the formation of cellular cGMP, thereby promoting polymerization of tubulin to microtubules and reversing abnormal bactericidal activity.

A selective impairment of NK-cell function in *bg/bg* mice prevents the initiation of antibody-dependent or antibody-independent cytolysis of tumor cells. The exact nature of the defect is not known, but it is believed to lie in the cell's lytic mechanism (Roder and Duwe, 1979). In vitro, these NK cells can be activated by treatment with interferon (Brunda et al., 1980). Defects have also been reported in cytotoxic T cells and macrophages (Mahoney et al., 1980; Saxena et al., 1982; Halle-Pannenko and Bruley-Rosset, 1985).

Beige homozygotes show a greater susceptibility to infection with pyogenic bacteria than do normal controls. This appears to be due to a defect in lysosomal function. The impairment is not of humoral origin because serum immunoglobulin levels are comparable in homozygotes, heterozygotes, and normal controls (Elin et al., 1974). Lane and Murphy (1972) demonstrated that the *bg* gene confers a greater susceptibility to spontaneous pneumonitis. In their study with SB/Le mice, which are homozygous for *bg*, *sa* (satin), and A^w (white-bellied agouti), and backcross offspring from an outcross to C57BL/6J-A^{w-J}, beige homozygotes (*sa bg/sa bg* and + *bg/sa bg*) had a significantly higher incidence of spontaneous pneumonitis than did nonbeige mice (*sa* +/*sa bg* and + +/*sa bg*). C57BL/6J-*bg^J/bg^J* and C3H/HeJ-*bg^{2J}/bg^{2J}* mice do not develop spontaneous pneumonitis under similar experimental conditions (J. B. Roths, The Jackson Laboratory, Bar Harbor, Maine, unpublished data).

C57BL/6 mice carrying the *bg* mutation or the pigment mutations pallid (*pa*), pearl (*pe*), light ear (*le*), pale ears (*ep*), maroon (*ru-2^{mr}*), or ruby eye (*ru*) have hypopigmentation, prolonged bleeding time, normal platelet numbers accompanied by reduced platelet granules, and decreased levels of platelet serotonin. These seven mutations map to separate chromosomal sites; however, bone marrow transplantion from normal C57BL/6 mice to irradiated mutants corrects the defects in platelet serotonin and bleeding time. This suggests that in each case there is a cellular basis to the deficiency (Novak et al., 1985; McGarry et al., 1986).

Experimentally, beige mice have played an important role in studies of hematopoietic differentiation. The giant lysosomal granules of beige mice provide an exquisite cytoplasmic marker for mast cells, PMNs, and osteoclasts in bone marrow chimeras (Murphy et al., 1973; Ash et al., 1980; Kitamura et al., 1981).

C3H/HeJ-*bg^{2J}/bg^{2J}* mice that survive to 17 months of age show a progressive neurological disorder accompanied by a nearly complete loss of cerebellar Purkinje cells (Murphy and Roths, 1978b). This lesion is less severe in similarly aged C57BL/6J-bg^J/bg^J mice.

The pigment mutations *ep, pa,* and *rp* (reduced pigmentation) affect lysosomal functions and lead to suppressed NK cell activity (Orn et al., 1982). An additional 29 pigment mutations have been described in mice, all of which are known to affect lysosomal biogenesis. However, little is known about the function of the immune system in these mice (Brandt et al., 1981).

Husbandry

Beige mice are more susceptible than immunocompetent mice to challenge by a wide variety of infectious agents, but they do not suffer from infections caused by indigenous microorganisms that are not pathogenic to immunocompetent mice. These animals survive well in a pathogen-free environment. Barrier isolation, as described in Chapter 4, is not required for the maintenance of these animals.

Reproduction

Beige mice of both sexes will breed.

db (Diabetes)

Genetics

The mutation diabetes (*db*) on chromosome 4 is a spontaneous autosomal recessive obesity gene discovered at the Jackson Laboratory, Bar Harbor, Maine, in the C57BL/KsJ inbred strain (Hummel et al., 1966). The gene has been transferred to the C57BL/6J and a number of other inbred strains. Several other alleles are present at the *db* locus: db^{2J} arose in an inbred brown (*b*) whirler (*wi*) stock (Lane, 1968), db^{3J} arose in strain 129/J (Leiter et al., 1980), and db^{Pas} arose in strain DW/J (Aubert et al., 1985). Another allele, db^{ad}, arose in a stock selected for large size and was called adipose (*ad*) (Falconer and Isaacson, 1959). In 1972 the gene was found to be allelic with *db* and was redesignated db^{ad} (Hummel et al., 1973).

Pathophysiology

Regardless of inbred strain background, *db/db* mice always exhibit a marked obesity associated with hyperphagia, hyperinsulinemia, and severe insulin resistance. However, ultimate development in *db/db* mice of a permanent diabetic condition (typified by chronic hyperglycemia, reduced serum insulin

A 129/J-db^{3J}/db^{3J} mouse (left) and its normal littermate (right). Mice homozygous for the diabetes mutation develop non-insulin-dependent diabetes and become obese. Photograph courtesy of the Jackson Laboratory, Bar Harbor, Maine.

levels, beta-cell necrosis, and pancreatic islet atrophy) depends entirely on the inbred strain background and gender (Coleman, 1978; Leiter et al., 1981). The mutation was named diabetes because the C57BL/KsJ strain of origin proved susceptible to its diabetogenic action. In this strain, *db/db* mice of both sexes develop an early-onset diabetes resembling in some respects the human non-insulin-dependent (type II) diabetes mellitus. On the contrary, the same mutation on the diabetes-resistant C57BL/6J inbred background did not produce beta-cell necrosis or permanent hyperglycemia; instead, the diabetogenic action of the mutation was well compensated by unrestricted hyperplasia of pancreatic beta cells and sustained hyperinsulinemia (Coleman, 1978).

Although type II diabetes in humans is not generally associated with an autoimmune etiology, the reports of both humoral and cell-mediated autoimmunity against pancreatic beta cells in C57BL/KsJ-*db/db* mice (Debray-Sachs et al., 1983), as well as the finding of immune complex deposition in the kidneys (Meade et al., 1981), suggested that this model might be an amalgam of features of type I (autoimmune) as well as type II diabetes in humans. An association between the *H-2* haplotype and diabetes susceptibility and resistance was initially suggested by a comparison of *db* gene expression on a variety of inbred strain backgrounds (Leiter et al., 1981). However, segregation analyses have shown that *H-2* does not control susceptibility; instead, male gender-linked factors, and possibly endogenous retroviral genes, appear to be the major modifiers of diabetes severity (Leiter, 1985; Leiter et al., 1987a). The question of the pathogenic role of the humoral and cell-mediated anti-beta-cell reactivities in *db/db* mice has also been resolved by combining this mutation with immunodeficiency genes such as severe combined immunodeficiency (*scid*) to compromise T- and B-lymphocyte function (Leiter et al., 1987b). This study showed that immunodeficient *db/db* mice still developed diabetes, indicating that autoimmunity is probably a reflection of islet cell destruction rather than its cause.

The C57BL/KsJ-*db/db* mouse has proven to be exceptionally useful as a model for analyzing the effects of chronic non-insulin-dependent diabetes on the immune system. There is early thymic involution and T lymphopenia from 8 weeks onward (Boillot et al., 1986); the disturbed metabolic milieu leads to depressed cell-mediated immunity and lymphokine production (Fernandes et al., 1978; Mandel and Mahmoud, 1978; Kazura et al., 1979; Pasko et al., 1981). Many of the T-cell functions that are impaired in vivo (for example, generation of alloreactive cells) are normal when assessed in vitro (Fernandes et al., 1978), which demonstrates the suppressive effect of the imbalanced metabolic environment in vivo. In contrast to the suppression of T-lymphocyte functions, certain B-lymphocyte functions are increased, including autoantibody production against islet cell cytoplasmic antigens, thymic hormones, and insulin (Dardenne et al., 1984; Serreze et al., 1988b; Yoon et al., 1988). Transfer of *db/db* marrow cells into lethally irradiated +/+ recipients rescued recipient mice from radiation death but did not transfer the diabetes syndrome (Leiter et al., 1987b).

Husbandry

Special procedures are not required to maintain these animals; however, their life span can be prolonged by dietary restriction, and the severity of the syndrome can be significantly diminished by feeding carbohydrate-free, protein-enriched defined diets (Leiter et al., 1983).

Reproduction

Diabetic mice of both sexes on all backgrounds will not mate, and females are hypogonadal. Therefore, breeding is accomplished with heterozygotes. To aid in the identification of heterozygous breeders on the C57BL/KsJ and C57BL/6J inbred backgrounds, the *db* gene has been placed in repulsion with the coat color mutation misty (*m*). Black, lean mice obtained from a cross of *db* +/+ *m* heterozygotes are used as breeders, whereas lean mice with grey coats (+ *m*/+ *m* genotype) are discarded. Black mice that become obese at weaning are the presumptive mutants (*db* +/*db* + genotype). When it is desirable to identify presumptive *db/db* mice as early as 3 days postpartum, breeding stocks are utlized in which *db* and *m* are maintained in coupling (*db m*/+ +). The mutant pups can be recognized by the absence of pigment in the paws and on the tip of the tail.

df (Ames Dwarf)

Genetics

Ames dwarf (*df*) appeared first in 1961 as an autosomal recessive mutation in a line of extreme nonagouti (a^e/a^e) mice derived from a cross between Goodale's giant and a pink-eyed stock (Schaible and Gowen, 1961). Although phenotypically similar to the mutation dwarf (*dw*) (see page 49), *df* is not allelic with *dw*. The *df* mutation, which maps to chromosome 11, is maintained on an outbred and on the NFR/N inbred backgrounds.

Pathophysiology

Homozygous *df* mice resemble *dw/dw* mice. Growth retardation is observed after 1 week of age, and by 2 months of age they are only one-half the weight of controls. The anterior pituitary lacks cells that produce either prolactin (Barkley et al., 1982) or growth hormone (Duquesnoy and Pedersen, 1981). Ames dwarf mice have a deficit of the T-cell component of the immune system, which becomes apparent by 3 weeks of age. Morphological abnormalities include depletion of lymphocytes in the thymus-dependent regions of lymph nodes and periarteriolar sheaths of the spleen follicles and a gradually progressive lymphopenia. This abnormality culminates in the development of progressive atrophy of lymphoid tissue, signs of wasting and infection, and early death. Other features of the T-cell deficit are depressed ability of spleen cells to induce graft-versus-host reaction, impaired ability to reject allogeneic skin grafts, and impaired phytohemagglutinin (PHA)-induced blastogenesis.

Both dwarf mutations have a similar deficit of T-cell function, although the deficit is more severe in *df* than in *dw* mice. There are subtle differences found in the kinetics of the immune response to sheep red blood cells (SRBCs). Although *dw* mice always develop more feeble plaque-forming cell (PFC) responses compared with control mice, they show an increase in PFC responses with increasing doses of SRBCs. In contrast, *df* mice do not show a dose-related dependency of immune response to SRBCs. The reason for this difference between the two dwarf mutations is not well understood, and numerous theoretical explanations have been postulated (Duquesnoy and Pedersen, 1981).

Husbandry

Mice carrying the *df* mutation should be maintained as described for those carrying *dw* (see page 49).

Reproduction

Females and nearly all male *df* homozygotes are sterile (Bartke, 1964); therefore, the stock must be maintained by breeding heterozygotes.

Dh (Dominant Hemimelia)

Genetics

The autosomal dominant mutant gene *Dh* arose spontaneously in a crossbred stock at the Institute of Animal Genetics, Edinburgh, Scotland (Carter, 1954). *Dh* is located on chromosome 1, approximately 3 centimorgans (cM) proximal to *Pep-3*. *Dh/+* mice are maintained on the BALB/cJBoy inbred and C57BL/6J × C3HeB/FeJLe-*a/a* F_1 hybrid backgrounds.

Pathophysiology

Both hetero- and homozygous *Dh* mice are congenitally asplenic and have multiple gastrointestinal, urogenital, and skeletal anomalies (polydactyly and oligodactyly) (Searle, 1959). Homozygotes, in which these anomalies are very severe, die shortly after birth. The abnormalities have been traced to a defect in the splanchnic mesoderm (Green, 1967).

Heterozygotes have enlarged lymph nodes and elevated numbers of circulating lymphocytes, granulocytes, and thrombocytes (Lozzio, 1972). They show decreased serum levels of IgM and IgG2 immunoglobulins but normal levels of IgG1. There is no evidence that *Dh* mice spontaneously produce autoantibodies (Fletcher et al., 1977). Both primary and secondary immune responses against SRBCs are reduced compared with those in eusplenic controls. Cell-mediated immunity, as measured by allograft rejection, is not impaired, and the response of lymph node cells to the mitogens concanavalin A (ConA), PHA, and lipopolysaccharide (LPS) is normal. Lymph nodes of *Dh/+* mice have a significantly reduced number of B cells, with a compensatory increase in the number of T cells. Hereditary asplenia is also associated with a significant reduction in the rate of T-cell maturation (helper T cells specifically) (Fletcher et al., 1977). The immunocompetence of *Dh* mice has been thoroughly reviewed by Welles and Battisto (1981).

Husbandry

Special husbandry procedures are not required for maintaining heterozygotes. Special husbandry procedures will not prolong the life of *Dh* homozygotes.

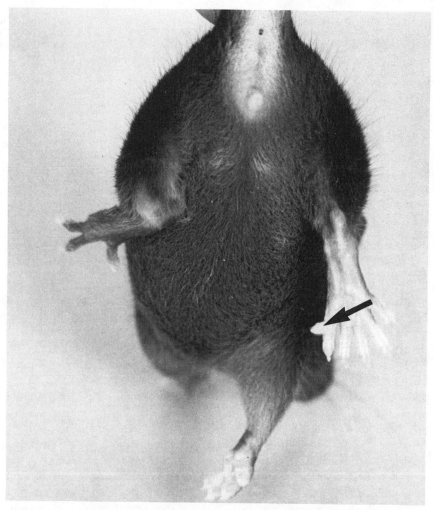

A B6C3/Fe-*a/a Dh/+* mouse. Mice carrying the gene dominant hemimelia commonly have polydactyly, as shown on the animal's right foot, which has six toes. Photograph courtesy of the Jackson Laboratory, Bar Harbor, Maine.

Reproduction

The *Dh* mutation is maintained by breeding heterozygous (*Dh/+*) females with homozygous normal (+/+) males or +/+ females with *Dh/+* males.

dw (Dwarf)

Genetics

The autosomal recessive mutation dwarf (*dw*) was discovered by Snell (1929) in a stock of mice obtained from an English mouse fancier. The mutation, which maps to chromosome 16, is now propagated in the segregating DW/J inbred strain. It has also been transferred to C3H/HeJ, C3H/HeN, C57BL/6J, and NFR/N backgrounds. C57BL/6J-*dw/dw* mice are preserved as frozen embryos. A point mutation producing dwarfed offspring on the C3H/HeJ background has been found to be a new allele, dw^J, at the dwarf locus (Eicher and Beamer, 1980).

Pathophysiology

Homozygous *dw* mice are one-fourth to one-third the size of normal mice. The anterior hypophysis, as in *df* mice, is deficient in acidophils and thyrotropic hormone-producing cells (Elftman and Wegelius, 1959). The immune system deficit was first described by Baroni and coworkers (Baroni, 1967; Baroni et al., 1967, 1969) and further characterized by Duquesnoy and coworkers (Duquesnoy and Good, 1970; Duquesnoy and Pedersen, 1981; Duquesnoy et al., 1970), who showed the deficiency to be in the T-cell system.

The thymus of *dw* homozygotes is characterized by atrophy and a marked

A DW/J-*dw/dw* mouse (left) and its normal littermate (right). Dwarf homozygotes are much smaller than control mice. Photograph courtesy of the Jackson Laboratory, Bar Harbor, Maine.

loss of lymphocytes in both the medulla and the cortex. A specific deficiency of lymphocytes in the cortex alone is observed in 25 percent of dwarf mice. Periodic acid-Schiff (PAS)-positive cells are prominent in the cortex. Peripheral lymph nodes are small, and the spleen has only one-tenth the normal number of lymphoid cells. Peripheral lymphoid tissues show hypocellularity, especially in the T-cell-dependent areas (paracortical areas of the lymph nodes and perifollicular sheaths in the spleen), but structurally they are similar to the lymphoid follicles of control mice. Dwarf mice have a marked reduction in peripheral blood lymphocytes but normal numbers of PMNs. Duquesnoy et al. (1970) noted that the immunologic deficiency becomes particularly evident after weaning at 21 days of age and showed that prolonged nursing delays its development. These observations raised the question of whether a factor present in the milk contributes to maintenance of the lymphoid system.

Dwarf mice have normal levels of serum IgG1, IgG2, IgM, and IgA. Hemagglutinating antibody titers to SRBCs (T cell dependent) and agglutinating antibody responses to *Brucella* antigen (T cell independent) are considerably lower in *dw/dw* mice than in controls. Spleen cells of homozygotes show a depressed capacity to initiate graft-versus-host reaction and a diminished response to PHA stimulation.

Pelletier et al. (1976) have found that dwarf mice have very low levels of serum thymic hormone activity (thymulin or facteur thymique serique) and minimal kidney disease. Littermate controls have normal levels of thymulin, and by 14 weeks they develop severe kidney lesions characterized by deposits of IgG1, IgG2, IgA, IgM, and C3 (a component of complement), which are found mainly in the glomerular mesangium. The serum of the heterozygous littermates contains antinuclear antibodies (ANAs) and anti-DNA autoantibodies, while dwarf homozygotes do not develop these autoantibodies. It has been postulated that *dw/+* mice have a precocious decrease of suppressor cells (Pelletier et al., 1976).

Schneider (1976) found normal delayed-type hypersensitivity and blastogenic responses in T cells of *dw* homozygotes, and Dumont et al. (1979) reported that *dw/dw* mice do not have major abnormalities in either T- or B-cell function. These findings, although contrary to most work done with these mice, demonstrate that the immunologic deficiencies described previously might not automatically accompany the dwarfism. In their recent review, Shultz and Sidman (1987) suggest that differences in genetic background and husbandry conditions might explain the wide differences seen in the immune competences of these mice. Recent studies report a mean life span in dwarf mice of 18 months (Schneider, 1976) to 26 months (Eicher and Beamer, 1980), compared with earlier studies reporting a life span of only 2–4 months (Baroni, 1967; Duquesnoy et al., 1970).

Husbandry

Investigators are encouraged to maintain homozygous *dw* mice in a pathogen-free environment because of the reported variability in levels of immune competence. Other special husbandry procedures that should be used because of the small size of these animals are to place food pellets in the bottom of the cages, as well as in feeders, and to use longer than normal sipper tubes for water.

Reproduction

Homozygous dwarf mice of both sexes are infertile. The mutation is propagated by matings of tested heterozygotes.

gl (Grey-Lethal)

Genetics

Grey-lethal (*gl*) is a recessive lethal mutation that arose spontaneously in a stock segregating for c^e (extreme dilution) (Grüneberg, 1935). It has been mapped to chromosome 10 (Lane, 1971). It is maintained on the inbred GL strain, whose genetic background has not been typed.

Pathophysiology

Grey-lethal is one of four osteopetrotic mutations in mice (see also *mi*, page 67; *oc*, page 73; and *op*, page 75). Animals have a grey coat, unless the mutation is carried on a homozygous nonagouti (*a/a*) background. The teeth of homozygotes fail to erupt, the animals develop severe osteopetrosis, and death occurs by 20–40 days of age (Grüneberg, 1936, 1938; Marks, 1984). Grüneberg (1938) reported a marked reduction in cortical thymocytes; however, since the teeth fail to erupt, it is likely that this thymic defect was caused by starvation. Wiktor-Jedrzejczak et al. (1983) observed normal thymic cellularity at 3 weeks of age and thymic involution, apparently associated with ill health, by 5 weeks of age.

The *gl/gl* mouse is anemic and has a reduced white cell count, although the granulocyte count is slightly elevated. Spleen cellularity is reduced; however, the number of splenic stem cells is increased (Wiktor-Jedrzejczak et al., 1981). Osteoclasts are small and reduced in number, and there appear to be two subpopulations, one with a normal acid phosphatase content and one with a reduced content (Marks and Walker, 1976). The rate of bone deposition is higher, the number of parafollicular (calcitonin-secreting) cells of the thyroid is greater (by eightfold), and the level of serum calcium is lower in *gl/gl* mice than in normal littermates. Parathyroid hormone administration does not cure the disease and only slightly raises serum calcium (Marks and Walker, 1969). The *gl/gl* mouse does not form normal bone marrow cavities in long bones. Responsiveness of splenic lymphocytes to T- and B-cell mitogens and of thymocytes to ConA is equal to or better than that of normal littermates, at least early in life (Wiktor-Jedrzejczak et al., 1981). The response of the *gl/gl* mouse to transplanted hematopoietic stem cells has not been completely characterized; however, cure can be effected by transplantation of normal spleen or bone marrow cells from histocompatible donors (Walker, 1975a,c). The defect can be transferred with spleen cells from *gl/gl* donors (Walker, 1975b). The *gl/gl* mouse might provide a model for studying the association between the immune system, especially T lymphocytes, and early events leading to osteopetrosis (Wiktor-Jedrzejczak et al., 1981).

Husbandry

Special husbandry procedures are not required for maintaining heterozygotes. Special husbandry procedures will not prolong the life of homozygotes.

Reproduction

The *gl* mutation is maintained by brother × sister mating of heterozygotes in a balanced stock. Litter sizes are normal, and there is no evidence for in utero deaths.

gld (Generalized Lymphoproliferative Disease)

Genetics

Generalized lymphoproliferative disease (*gld*) is an autosomal recessive mutation that occurred spontaneously in the C3H/HeJ strain (Murphy et al., 1982). The gene determines the development of early-onset lymphoid hyperplasia with autoimmunity. Although phenotypically similar to C3H/HeJ-lpr/lpr, early tests indicated that *gld* and *lpr* were not allelic. The *gld* mutation has been mapped to chromosome 1 between *Pep-3* and *Lp*. It has been transferred to inbred strains C57BL/6J and SJL/J by multiple (more than eight) cross–intercross cycles of matings.

HEREDITARY IMMUNODEFICIENCIES   53

A C3H/HeJ-*gld/gld* mouse. Mice affected with generalized lymphoproliferative disease develop enlarged lymph nodes, as shown by this animal's enlarged prescapular lymph node. Photograph courtesy of the Jackson Laboratory, Bar Harbor, Maine.

Pathophysiology

The mean life span of C3H/HeJ-*gld/gld* mice is 53 weeks for females and 57 weeks for males. They appear healthy, except for enlarged lymph nodes, and are active until several weeks prior to death.

Significant lymph node enlargement is apparent as early as 12 weeks of age, and by 20 weeks, lymph nodes are 60-fold heavier than those in coisogeneic C3H/HeJ controls. A major regression of lymph node mass occurs at about 30 weeks of age (reduction from 60- to 25-fold heavier), followed by a second proliferative phase in which the lymph node mass becomes 96-fold heavier than normal by 47 weeks of age.

Histological analysis reveals a loss of nodular architecture with a predominant proliferation of lymphocytes, including an admixture of plasma cells and histiocytes. Fibrosis and multinucleated giant cells are frequently observed in *gld* mice that reach more than 40 weeks of age (Roths et al., 1984).

Fluorescence-activated cell sorting (FACS) analysis of enlarged lymph nodes from C3H/HeJ-*gld/gld* mice reveals an aberrant population of Thy-1$^+$, Ly-1$^+$, L3T4$^-$, and Lyt-2$^-$ T cells that also bear the B-cell marker Ly-5 but not surface immunoglobulin (sIg) or Ia (Davidson et al., 1985). DNA probe analysis for the immunoglobulin heavy chain and the T-cell antigen receptor beta-chain rearrangements demonstrates that abnormal cells found in lymph nodes of *gld* homozygotes are of T-cell origin (Davidson et al.,

1986). Ishida et al. (1987) postulate, based on investigations comparing normal and *gld* mutant mice, that the aberrant cell population in *gld* mice is a normal regulatory T cell that is proliferating abnormally. Lymphocytes of *gld* mutant mice are deficient in their ability to produce IL-2 and have reduced proliferative responses to alloantigens in mixed lymphocyte reactions (MLRs) (Davidson et al., 1985). However, clones of T cells with a normal surface phenotype that are established from the lymph nodes of *gld* homozygotes are fully capable of recognizing class I molecules on target cells and functioning as cytotoxic lymphocytes (Yui et al., 1987). Splenomegaly, in which the spleens become eight times heavier than those of controls by 47 weeks of age, is primarily caused by the expansion of white pulp. Neonatal thymectomy prevents the lymphadenopathy in *gld/gld* mice, thus implicating a major role for thymic-derived lymphocytes in the pathogenesis of this disease.

Only 14 percent of autopsied C3H/HeJ-*gld/gld* homozygotes have significant lupuslike renal disease. Immune complexes are found in the kidneys (primarily the mesangium) of all adult C3H/HeJ-*gld/gld* mice. Histological analysis of kidneys of the majority of *gld/gld* mice, however, reveals only a minor focal renal pathology.

Serologically, C3H/HeJ-*gld/gld* mice develop antinuclear antibodies, including anti-double-stranded DNA (anti-dsDNA), thymus-binding autoantibody, and hyperimmunoglobulinemia with major increases in several isotypes. Recent evidence suggests that expression of the *xid* gene can decrease the B-cell manifestations of autoimmunity without affecting the abnormal T-cell expansion (Seldin et al., 1987). Interstitial pneumonitis is found in virtually all *gld* mice and is a likely cause of morbidity (Roths et al., 1984).

Husbandry

Special husbandry procedures are not required.

Reproduction

Productivity of *gld* homozygotes is similar to that of their normal (+/+) counterparts.

Hc^0 (Hemolytic Complement Absent)

Genetics

Hemolytic complement (*Hc*) is located on chromosome 2. Two alleles are known: Hc^1 determines the presence of the fifth component of complement (C5) in serum; Hc^0 determines its absence (Erickson et al., 1964). Heterozygotes have half the amount of complement as do homozygous Hc^1 mice.

Pathophysiology

Lack of complement activity was discovered by Rosenberg and Tachibana (1962) in two strains of mice, DBA/2 and B10.D2/oSnJ. In the early literature, C5 was called beta-globulin, MuB1 (Cinader and Dubiski, 1964), or β1F glycoprotein (Nilsson and Müller-Eberhard, 1965). In humans, total C5 deficiency corresponds to a genetic locus linked to the major histocompatibility complex (Rosenfeld et al., 1977).

Studies by Patel and Minta (1979) demonstrated the incorporation of ^{14}C-labeled amino acids into a C5-like protein by C5-deficient cells in vitro. The protein elaborated by C5-deficient macrophages was the 200-kDa pro-C5 molecule deficient in carbohydrate and not secreted (Ooi and Colten, 1979). The absence of circulating C5 prohibits the formation of the membrane attack complex (C5b-9) during complement activation.

Strains of mice that bear the Hc^0 allele have low serum hemolytic complement activity only because of a deficiency of C5. There are no circulating complement inhibitors (Nilsson and Müller-Eberhard, 1967). When strains of mice carrying the homozygous Hc^0 allele (complement deficient) are compared with strains homozygous for the Hc^1 allele, they appear to be more susceptible to a variety of infectious agents inoculated experimentally (Caren and Rosenberg, 1966; Glynn and Medhurst, 1967; Morelli and Rosenberg, 1971; Easmon and Glynn, 1976; Hicks et al., 1978). Congenic strains differing only by the *Hc* allele have been used to demonstrate delayed pulmonary clearance of *Staphylococcus aureus* in C5-deficient mice (Cerquetti et al., 1983). There is little evidence, however, that C5-deficient mice are more susceptible to spontaneous infectious diseases.

Husbandry

Special husbandry procedures are not required based on the presence of Hc^0 alone. However, specific-pathogen-free conditions are recommended for certain strains bearing Hc^0 (e.g., DBA/2) that appear to be extremely susceptible to viral pathogens.

Reproduction

Most Hc^0 strains reproduce normally.

hr (Hairless); *hr^{rh}* (Rhino)

Genetics

The original recessive mutation to *hr*, located on chromosome 14, was discovered in a wild mouse captured in a London aviary in 1924 (Brooke,

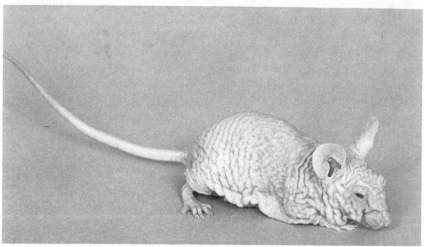

HRS/J-*hr/hr* (top) and RHJ/Le-*hr^rh/hr^rh* (bottom) mice. Hairless homozygotes are born with pelage but lose their hair after 10 days. Rhino homozygotes are hairless and have excessive skin folding. Photographs courtesy of the Jackson Laboratory, Bar Harbor, Maine.

1926). It has recently been reported that the mutation results from the integration of a provirus into the normal allele at the hairless locus (Stoye et al., 1988). Studies of abnormalities of the immune system in *hr/hr* mice have focused on the HRS/J strain, which originated from a cross between an outbred stock carrying the *hr* mutation and the BALB/cGn strain (Heiniger et al., 1974). Homozygosity for a second allele at the *hr* locus, called rhino (hr^{rh}), also causes immunologic dysfunction.

Pathophysiology

Unlike nude mice, which are hairless throughout life, homozygotes for mutant alleles at the *hr* locus develop a normal pelage up to 10 days of age, at which time they lose their hair. Although a few thin hairs grow in at monthly intervals, the mice are essentially hairless by 5–6 weeks of age. Alopecia in these mice appears to be related to the malpositioning of the internal root sheath. The most striking feature of the skin following hair loss is the formation of cutaneous cysts from hair follicles and isolated sebaceous glands (Mann, 1971). Investigations of the immune system of hairless mice were stimulated by the finding that they develop a high incidence of thymic lymphomas. HRS/J-*hr/hr* mice show a 45 percent incidence of thymic lymphoma by 10 months of age; HRS/J-*hr/*+ mice show a 1 percent incidence (Meier et al., 1969). Although hairless mice are euthymic, by 6 months of age there is marked thymic cortical atrophy (Heiniger et al., 1974). Defective immune function in hairless mice is associated with T-cell abnormalities. Splenic T cells from this mutant have a depressed proliferative response to I-region alloantigens (Morrissey et al., 1980). In addition, spleens show an inversion in the normal proportions of Ly-1^+ and Ly-123^+ T cells (Reske-Kunz et al., 1979). Reduced humoral immune responses to thymic-dependent antigens have also been reported in these mice (Heiniger et al., 1974). An association between defective immune function and lymphomagenesis in hairless mice is suggested by the reduced immune responsiveness of this mutant to syngeneic lymphoma cells and to purified murine leukemia viruses (MuLVs) (Johnson and Meier, 1981).

HRS/J-*hr/hr* mice and their heterozygous littermates have been found to be unusually susceptible to infection with *Listeria monocytogenes* (Archinal and Wilder, 1988a). During studies on oxidative metabolic activities associated with the respiratory burst in mutant and normal mice, the liberation of superoxide anion and the chemiluminescence response of macrophages were both significantly diminished in the caseinate-elicited cells from *hr/hr* and *hr/*+ mice (Archinal and Wilder, 1988b). These results are compatible with the hypothesis that the defect in resistance to *L. monocytogenes* in mutant mice is caused by an intrinsic deficiency in the mobilization and the listericidal activity of the macrophage.

Homozygosity for a second deleterious allele at the *hr* locus, called rhino (hr^{rh}), also results in the loss of hair and reduced responsiveness to thymic-dependent antigens. Rhino mice are similar in appearance to hairless mice, except that their skin becomes thickened and markedly wrinkled (Mann, 1971). As with hairless mice, decreased humoral immune responses to thymic-dependent antigens in rhino mice appear to be due to reduced helper T-cell function (Takaoki and Kawaji, 1980).

Husbandry

Special husbandry procedures are not required to maintain hairless and rhino mice, which, with the exception of occasional skin abscesses, show no increased susceptibility to infection.

Reproduction

Hairless and rhino mice are fertile, but females do not nurse their young well. The most effective breeding system uses homozygous mutant males and heterozygous females.

lh (Lethargic)

Genetics

The recessive mutation lethargic (*lh*), first described by Dickie (1964), occurred spontaneously at the Jackson Laboratory, Bar Harbor, Maine, in 1962 in strain BALB/cGn. The *lh* locus is approximately 24 cM from the centromere on chromosome 2.

Pathophysiology

Homozygous *lh* mice demonstrate an instability of gait by 15 days of age, and the majority (68–80 percent) die by 45 days. The pleiotropic effects of *lh* include an absence of subcutaneous fat and abnormalities of the lymphoid system. At 4 weeks of age *lh* mice are less than 50 percent of the weight of their normal littermates. The neuropathologic phenotype has been reviewed by Sidman et al. (1965).

Lymphoid organs of *lh/lh* mice, including thymus, spleen, lymph nodes, and Peyer's patches, are smaller than those of normal littermates. Lethargic mice have reduced erythrocyte packed-cell volumes (36 percent) and severe leukocytopenia. Homozygous *lh* mice have defects in cell-mediated immune functions. For example, the ability to reject skin allografts is defective in 30-day-old homozygotes but is normal in 40-day-old homozygotes. This transitional defect in cell-mediated immunity also is found in the ability to induce graft-versus-host reactions. Serum IgG1 and IgG2a levels are higher in mutant mice than they are in controls. Extensive studies by Dung and colleagues on the anatomic, neurologic, hematologic, and immunologic aspects of the lethargic mutation have been reviewed (Dung, 1981).

Husbandry

Compared with controls, *lh/lh* mice have a reduced capacity to resist hypothermia when exposed to extremely cold (4°C) ambient temperature. Even under conventional mouse room conditions, young (less than 45-day-old) *lh/lh* mice are hypothermic (Dung, 1981).

Reproduction

Homozygous survivors of both sexes can breed, but their reproductivity is low (Green, 1981). At the Jackson Laboratory, Bar Harbor, Maine, this mutation is maintained by continued backcrosses to C57BL/6J × C3H/HeSnJ F1 hybrid mice (see Chapter 5).

lpr (Lymphoproliferation)

Genetics

The mutant gene lymphoproliferation (*lpr*) is a single autosomal recessive gene that arose spontaneously in the twelfth generation of inbreeding of the developing MRL/Mp strain (Murphy and Roths, 1977). The origin of this strain and several of its sublines has been reviewed (Roths, 1987). Despite numerous attempts to demonstrate linkage, the location of the *lpr* gene is unknown. The gene has been transferred by multiple cross–intercross matings to inbred strains AKR/J, BALB/cBy, C3H/HeJ, C57BL/6J, C57BL/10Sn, and SJL/J.

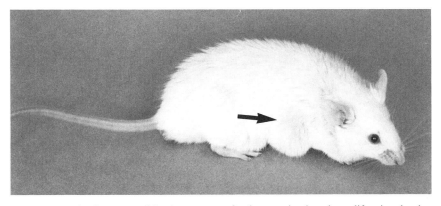

An MRL/MpJ-*lpr/lpr* mouse. Mice homozygous for the mutation lymphoproliferation develop enlarged lymph nodes, as shown by this animal's enlarged prescapular node. Photograph courtesy of the Jackson Laboratory, Bar Harbor, Maine.

Pathophysiology

The mean life span of MRL/Mp-*lpr/lpr* females and males is 17 and 22 weeks, respectively. Homozygotes die with massive generalized enlargement of the lymph nodes. Subcutaneous lymph nodes can be palpated as early as 10 weeks of age (Murphy and Roths, 1978a).

Enlarged lymph nodes of homozygotes are characterized predominantly by lymphocytic proliferation with an admixture of histiocytes, plasma cells, and immunoblasts. There is also blurring of the nodal architecture. Multiple attempts to transplant the lymphoid mass to congeneic normal recipients have been unsuccessful. The *lpr*-induced lymphoproliferation is presumed to be hyperplastic, not neoplastic. Lymphoproliferation in 4-month-old MRL/Mp-*lpr/lpr* mice is composed of approximately 90 percent T cells. Although the frequency of B cells is reduced to 7 percent from the normal (+/+) level of 28 percent, there is an absolute increase in the numbers of B cells at this age. Definitive FACS analyses employing monoclonal antibodies indicate that hyperplastic lymph node cells stain for Thy-1 (dull), Ly-1 (dull), H-11, and Ly-5. Such cells are negative for L3T4, Ly-2, sIg, ThB, and Ia-associated antigens. Thus, *lpr/lpr* mice have abnormal populations of T cells that express the Ly-5 marker usually confined to B cells (Morse et al., 1982).

Treatment of MRL-*lpr/lpr* mice from birth to 11 weeks of age with the monoclonal antibody Mel-14 results in a 10- to 20-fold reduction in lymphadenopathy and induces marked splenomegaly. Since Mel-14 recognizes the lymphocyte surface receptor gp90, which is responsible for lymph node homing, it is speculated that lymphadenopathy occurs in MRL-*lpr/lpr* mice because of increased homing of gp90$^+$ T cells to the lymph nodes (Mountz et al., 1988).

The spleen weight of MRL-*lpr/lpr* mice is sevenfold larger than that of controls. While lymphoid tissue in the area of the thymus appears enlarged in moribund mice, most of the mass is due to enlarged lymph nodes rather than to an enlarged thymus.

MRL/Mp-*lpr/lpr* mice develop immune complex glomerulonephritis. Generalized arterial disease, characterized by polyarteritis and degenerative arteriolar lesions without cellular inflammation, is also found. The lungs of *lpr/lpr* mice show extensive perivascular and peribronchial lymphocytic infiltration with only occasional atelectasis and exudate. Severe bronchopneumonia can occur, but interstitial pneumonitis is not characteristic (Murphy and Roths, 1978a). Joints of MRL/Mp-*lpr/lpr* mice show arthritic changes resembling those of rheumatoid arthritis (Andrews et al., 1978). Coronary artery disease and myocardial infarction can be a contributing cause of death (Hang et al., 1982). Near the end of their life span, MRL/Mp-*lpr/lpr* mice develop erythematous skin lesions, necrosis of the ears, swollen feet, and, frequently, generalized edema (Murphy and Roths, 1978a).

Serologically, *lpr/lpr* mice have a fivefold increase in gamma-region proteins. Twofold increases in IgA, IgM, and IgG2b and sixfold increases in IgG1 and IgG2a have been identified by radial immunodiffusion. Sera of *lpr/lpr* mice are uniformly positive for antinuclear autoantibodies by 12 weeks of age. The presence of anti-single-stranded DNA (anti-ssDNA), anti-dsDNA, and anti-Smith (anti-Sm) autoantibodies is characteristic. Autoantibodies directed against thymocyte or erythrocyte surface antigens are detected infrequently (Murphy and Roths, 1978a). Retroviral gp70 immune complexes and IgM and IgG rheumatoid factors are found in significant concentrations in *lpr/lpr* mice (Andrews et al., 1978).

Muraoka and Miller (1988) found that MRL-*lpr/lpr* and C57BL/6-*lpr/lpr* mice, but not MRL/Mp or C57BL/6 mice, develop lymph node and bone marrow-derived cytotoxic lymphocytes directed against self (H-2^k) determinants in a ^{51}Cr release assay with mitogen-induced lymphoblasts. They concluded that at least some of the self-reactive cells are generated as part of the *lpr* defect.

Both the longevity and pathologic sequelae associated with the *lpr* gene depend on the background of the host. Life spans of female *lpr* homozygotes on the AKR/J, C3H/HeJ, and C57BL/6J inbred strains are between 42 and 52 weeks. SJL/J-*lpr/lpr* homozygotes have life spans similar to those of MRL/Mp-*lpr/lpr* mice (17–22 weeks) (Roths, 1987).

The degree of lymphoproliferation varies greatly among strains. At 26 weeks of age the mean weight of the lymph nodes is 20-fold greater in AKR/J-*lpr/lpr* mice than in congeneic normal AKR/J mice. This compares with a 116-fold increase over control lymph node weight in C3H/HeJ-*lpr/lpr* mice (Morse et al., 1985). The rank order of increasing lymphoid hyperplasia for five congenic strains is AKR/J-*lpr/lpr* < C57BL/6J-*lpr/lpr* < SJL/J-*lpr/lpr* < MRL/Mp-*lpr/lpr* < C3H/HeJ-*lpr/lpr*.

Genetic background also affects spontaneous production of anti-dsDNA autoantibodies. At 6 months of age the percentage binding of radiolabeled dsDNA in *lpr/lpr* mice has been found to be 5 percent on C57BL/6J, 20 percent on AKR/J, 26 percent on C3H/HeJ, and 49 percent on MRL/Mp (Izui et al., 1984). Renal pathology is extensive in MRL/Mp-*lpr/lpr* mice at 4–7 months of age, compared with negligible kidney disease in C57BL/6J-*lpr/lpr* and C3H/HeJ-*lpr/lpr* mice as old as 14–16 months of age. AKR/J-*lpr/lpr* mice have mild renal disease. Levels of urinary protein parallel the degree of renal histopathology (Kelley and Roths, 1985).

Husbandry

Special husbandry procedures are not required for maintaining MRL/Mp-*lpr/lpr*, C57BL/6J-*lpr/lpr*, or C3H/HeJ-*lpr/lpr* mice. The severity of disease and early morbidity make it extremely difficult to maintain a colony of SJL/J-*lpr/lpr* mice.

Reproduction

MRL/Mp-+/+ and MRL/Mp-*lpr/lpr* are independently inbred with periodic crosses (every 5–10 inbreeding generations) to the MRL/Mp-+/+ reference strain. MRL/Mp mice are large and docile; males rarely fight. MRL/Mp-*lpr/lpr* breeders produce an average of seven pups per litter, with 96 percent of the offspring surviving to weaning. Because of the early mortality of both males and females, however, maintaining a colony of MRL-*lpr/lpr* breeders requires careful management. An average of only two litters per breeding female should be expected. Approximately one-tenth of the breeding colony should be retired and replaced by newly weaned breeders each week. Trio matings are successful but require breeding boxes adequate to house the increased population of preweanling mice.

Lps^d (Lipopolysaccharide Response, Defective)

Genetics

The *Lps* locus in mice controls a number of dominant or codominant traits that are elicited in response to the complex glycolipid component (endotoxin) of the outer membrane of gram-negative bacteria (Mergenhagen and Pluznik, 1984). By common usage, the term lipopolysaccharide (LPS) or lipid A is considered to be synonymous with this endotoxic component. The *Lps* gene is located on chromosome 4 (Watson et al., 1978a). Mice homozygous for the Lps^n locus have normal in vivo and in vitro responses to endotoxin, while mice carrying the Lps^d allele have defective responses. The first documented allele for defective LPS responses occurred in the C3H/HeJ strain between 1960 and 1968, presumably by mutation (Glode and Rosenstreich, 1976; Watson and Riblet, 1974). Subsequently, additional—perhaps identical—defective mutations at the *Lps* locus have been detected in C57BL/10ScCr and C57BL/10ScN strains (Green, 1981; Morrison and Ryan, 1979; Vogel et al., 1981).

Pathophysiology

C3H/HeJ mice, which have the allele Lps^d, show a 20- to 38-fold increase in resistance to endotoxin lethality over the highly susceptible A/HeJ strain (Sultzer, 1968). At the same time they show a greater susceptibility to *Salmonella typhimurium* administered systemically than do strains bearing the

Lps^n allele (O'Brien et al., 1985). In addition, the initial clearance of both *S. typhimurium* and *Escherichia coli* from mucosal surfaces is significantly impaired in Lps^d mice (Edén et al., 1988). The increased susceptibility to *S. typhimurium* and *E. coli* regulated by the Lps^d locus is independent of the expression of the Ity^s locus in mice (see page 36). Spleen cells from the Lps^d strain C3H/HeJ cannot support LPS-induced herpes simplex virus (HSV) replication, and the C3H/HeJ strain is resistant to lethal challenge with HSV (Kirchner et al., 1978). The dichotomy seen between resistance to endotoxin and HSV and susceptibility to *S. typhimurium* in C3H/HeJ is reflected in the numerous abnormal responses to LPS conferred by Lps^d. To date, these abnormalities have been studied best in the immunologic and reticuloendothelial systems.

Watson and Riblet (1975) found that B lymphocytes from C3H/HeJ mice were unable to proliferate in the presence of LPS. This defect was shown to reside in a single codominantly expressed autosomal gene (Coutinho et al., 1975). In addition to these defective mitogenic and polyclonal PFC responses, Skidmore et al. (1976) demonstrated that C3H/HeJ mice were deficient in the LPS-induced adjuvant response to antibody production. These investigators also demonstrated that the normal abrogation of tolerance to human IgG associated with LPS was not observed in the Lps^d strains. The LPS-induced phenotypic differentiation of B cells (immunoglobulin, Ia expression) and T cells (Thy-1 expression) was shown to be lacking in Lps^d strains (Koenig et al., 1977; Watson, 1977). Taken collectively, these observations suggest a significant role for the Lps^n gene product in the response of B cells (and some T cells) to LPS.

Glode et al. (1977) found that macrophages collected from C3H/HeJ mice were intrinsically unresponsive to the cytotoxic effects of endotoxin. Macrophages from C3H/HeJ mice have no LPS-induced inhibition of phagocytosis and produce very low levels of IL-1 and prostaglandin E_2 on exposure to LPS (Vogel et al., 1979). Chedid et al. (1976) reported that macrophages collected from Lps^d strains of mice were not activated by endotoxin to kill tumor cells. Backcross linkage analysis conducted between Lps^d and Lps^n strains demonstrated complete concordance in the expression of macrophage tumoricidal capacity and B-cell proliferative responses (Ruco et al., 1978). A wide variety of additional abnormal biological responses are conferred by Lps^d. Augmentation of neutrophil migration in response to LPS administered intraperitoneally has been described (Sultzer, 1968; Moeller et al., 1978). Lps^d strains fail to make type I interferon (α/β) in vivo (Apte et al., 1977), and Lps^d fibroblasts have decreased glucose utilization in response to LPS in vitro (Ryan and McAdam, 1977). Using backcross linkage analysis involving the C3H/HeJ and C57BL/6J (Lps^n/Lps^n) strains, Watson et al. (1978b) observed that Lps^d mice failed to produce serum amyloid A (SAA) and granulocyte-macrophage colony-stimulating factor in response to LPS. They also discovered that Lps^d mice develop hypothermic responses following LPS injection.

The low responsiveness of C3H/HeJ mice to lipid A is the result of an inability to activate target cells rather than an inability of cells to respond to endogenous mediators (Mergenhagen and Pluznik, 1984). Selective mitogenic responses have been elicited in Lps^d mice by using detergent-dissociated fractions of LPS (Vukajlovich and Morrison, 1984). Vogel et al. (1984) demonstrated that lipid A defective in 3-deoxy-D-mannooctulosonic acid as well as ester-linked lauryl and myristoyl residues is capable of inducing many of the biological effects of intact lipid A in both Lps^n- and Lps^d-containing macrophages. They speculate that the defect in C3H/HeJ mouse cells associated with the Lps^d gene may be related to the processing of lipid A to a suitable stimulatory form. Although the nature of LPS cell binding and activation is not completely known, it appears that in Lps^d macrophages, endotoxin fails to produce protein phosphorylation and, thus, complete signal transduction (Prpic et al., 1987).

Husbandry

Special husbandry procedures are not required.

Reproduction

These animals reproduce normally.

me (Motheaten); me^v (Viable Motheaten)

Genetics

The recessive mutation motheaten (*me*) occurred spontaneously in 1965 in the C57BL/6J strain at the Jackson Laboratory, Bar Harbor, Maine (Dickie, 1967). The *me* locus has been mapped to chromosome 6, 21.9 cM distal to Mi^{wh}. A second spontaneous mutation at the *me* locus occurred in C57BL/6J mice at the Jackson Laboratory in 1980 (Shultz et al., 1984). This mutant allele was designated viable motheaten (me^v) because of its increased longevity. In addition to the strain of origin, motheaten is available on strains C3H/HeN, C3HeB/FeJLe, C57BL/6N, NFS/N, NZB/N, and P/N. Viable motheaten has been partially or fully backcrossed onto several inbred strains, including AKR/J, C3H/HeJ, DBA/2J, MRL/Mp, and SJL/J.

A C3HeB/FeJLe-*a/a me/me* mouse. Motheaten homozygotes have patchy absence of hair, are small, and develop terminal pneumonia (hunched position). Photograph courtesy of the Jackson Laboratory, Bar Harbor, Maine.

Pathophysiology

C57BL/6J-*me/me* mice are recognized by 3–4 days of age by neutrophilic aggregates in the skin. Later, subepidermal accumulations of neutrophils displace hair follicles and result in the patchy absence of pigment. Motheaten mice are considerably smaller than normal littermates from 4 days of age onward.

C57BL/6J-*me/me* mice have the shortest life span of any mouse strain with a single mutation affecting the immune system. Only one-fifth of these homozygotes survive to weaning, and none survive longer than 8 weeks. The mean longevity of C57BL/6J-*me/me* and C57BL/6-*me^v/me^v* mice is 3.1 and 8.7 weeks, respectively (Green and Shultz, 1975; Shultz et al., 1984). However, the average life span of other inbred strains carrying the *me* locus can be as long as 6–8 weeks (C. T. Hansen, Division of Research Services, National Institutes of Health, unpublished data). The immediate cause of death appears to be pneumonia. Large numbers of macrophages, which often contain fine crystalline material associated with extravasation of erythrocytes, are found in the alveoli.

Until approximately 3 weeks of age, the thymuses of motheaten mice appear to be histologically normal, although they are smaller in size than those of controls. Thereafter, the thymuses show marked cortical involution and are absent in the oldest survivors. Splenomegaly, characterized by an increase in erythropoiesis and myelopoiesis and a loss of white pulp, is seen in both *me* and *me^v* homozygotes. Lymph nodes might be slightly enlarged, but follicles are absent. Large numbers of atypical plasma cells with Russell bodies are found in *me^v/me^v* mice. Peyer's patches are reduced in size and have no follicles. Bone marrow shows reduced erythropoiesis and increased myelopoiesis. The peripheral blood is characterized by an increased number of neutrophils and a decreased number of lymphocytes. In the oldest *me* and *me^v* homozygotes, renal disease is evident. By 20 weeks of age, *me^v/me^v* mice have marked glomerulonephritis with an associated increase in blood urea nitrogen. Male *me^v/me^v* mice are sterile because of depletion of Leydig's cells, lowered testosterone levels, and impaired spermatogenesis.

Motheaten mice have severe deficiencies in both humoral and cellular immunity. Homozygous *me* mice have increased levels of all major classes of immunoglobulins, and IgM levels in particular are up to 25 times higher in 6-week-old *me/me* mice than they are in controls. Viable motheaten mice have increased numbers of Ly-1$^+$ B cells (Sidman et al., 1986b), impaired proliferative responses to B-cell mitogens (Sidman et al., 1978), and polyclonal B-cell activation (Sidman et al., 1978). Although numbers of T lymphocytes are normal, *me/me* mice have impaired proliferative responses to T-cell mitogens and lack the capacity to develop cytotoxic killer cells against allogeneic cells. Greiner et al. (1986a) have postulated that a subset of TdT$^+$ bone marrow cells are almost totally depleted in *me* homozygotes, and prothymocytes appear to be developmentally arrested at the pre-TdT$^+$-cell stage. It has been postulated that the maturation of prothymocytes in *me^v/me^v* mice is arrested because of a defect in the radiosensitive compartment of the bone marrow microenvironment (Komschlies et al., 1987). There is an early onset of autoimmunity characterized by the presence of autoantibodies against thymocytes and dsDNA (Shultz and Zurier, 1978). Immune complexes (granular deposition of IgM and IgG) have been identified in renal glomeruli by immunofluorescence microscopy (Shultz and Green, 1976). Sherr et al. (1987) postulated that the increased population of Ly-1$^+$ B cells has a direct role in the autoimmune disease of *me^v/me^v* mice by secreting auto(idiotype)-reactive antibodies and elaborating helper lymphokines that drive autoreactive subsets to secretion.

Painter et al. (1988) evaluated the antibodies produced by several hundred hybridomas derived from C57BL/6JSmn-*me^v/me^v* mice. Of 33 immunoglobulin-secreting hybridomas, 17 exhibited reactivity to autoantigens. While some of the monoclonal antibodies (mAbs) recognized a single autoantigen, others recognized multiple antigenic specificities. Northern blotting analysis for V-gene families in the *me^v*-derived autoantibodies was interpreted as showing random V_H family selection but biased use of V_κ gene families. They concluded that the immunoregulatory defect in *me^v* mutants operated at a more generalized level than at the V_H or V_κ loci.

Using Whitlock-Witte cultures for lymphocytes and Dexler cultures for myeloid cells, Hayashi et al. (1988) found that neither lymphocytes nor granulocytes were formed from mutant me^v/me^v bone marrow cells alone or together with normal cells. There was no evidence for soluble mediators of suppression. Thus, the me^v/me^v mouse might be a good model for investigating cell-associated molecules that normally limit progenitor cell replication.

Husbandry

Special husbandry procedures are not required for maintaining *me* and me^v mice. Life span is not increased by maintaining these animals under germfree conditions (Lutzner and Hansen, 1976).

Reproduction

Propagation of *me* and me^v mice is difficult, but can be done by breeding heterozygotes (*me*/+) or by transplanting ovaries of C57BL/6J-*me/me* females to histocompatible hosts with subsequent mating to C57BL/6J males to produce known heterozygotes (see Chapter 5).

mi (Microphthalmia)

Genetics

Microphthalmia (*mi*) is a semidominant gene that arose in the descendents of an irradiated male (Hertwig, 1942). Numerous alleles occur at this locus,

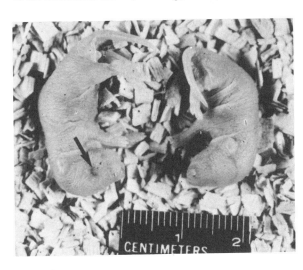

A mouse carrying the mutation microphthalmia (*mi*) and its normal littermate. Mutants of microphthalmic stock (right) can be identified at birth by the absence of retinal pigment, which is shown with an arrow on the normal littermate (left).

which has been mapped to chromosome 6, and some combinations show interallelic complementation (Green, 1981). The gene is maintained on C57BL/6 and CBA/H-T6 backgrounds.

Pathophysiology

Heterozygous *mi* mice often have white spotting on the head, tail, and belly and exhibit reduced iris pigment. Homozygotes are devoid of eye and coat pigments, have small eyes, and lack incisors. Microphthalmic mice (like *gl/gl*, *oc/oc*, and *op/op* mice) exhibit osteopetrosis. Most homozygotes die at about the weaning age or a little older.

Osteoclasts in *mi/mi* mice are small and numerous, have fewer nuclei, lack cytoplasmic vacuoles, and have rudimentary ruffled borders (Marks and Walker, 1981). Like *gl/gl* mice, *mi/mi* mice appear to have two populations of osteoclasts, one with normal and one with reduced acid phosphatase (Marks and Walker, 1976). Homozygotes have wide expanses of osteoid, unmineralized bone matrix. Osteoblasts are numerous and active (Marks and Walker, 1969). Like the other osteopetrotic mutants, *mi/mi* mice show hyperplasia of the calcitonin-secreting cells of the thyroid, hypocalcemia, and hypophosphatemia (Marks and Walker, 1969). Serum levels of 1,25 dihydroxyvitamin D are elevated, and levels of the 24,25 metabolite of vitamin D are reduced (Zerwekh et al., 1987).

The *mi/mi* mouse is anemic and has a leukocyte count in the low normal range. Spleen stem cell populations are markedly increased, and splenomegaly is present. Marrow hemopoietic stem cell populations are normal (Wiktor-Jedrzejczak et al., 1981).

Studies using cell isolates from whole spleens have demonstrated defects in monocyte (Chambers and Loutit, 1979; Minkin, 1981), macrophage (Chambers and Loutit, 1979; Minkin, 1981), and lymphocyte (Olsen et al., 1978; Minkin et al., 1982) function; however, Schneider and Marks (1983), using ficoll-hypaque separated splenic lymphoid cells, did not observe reduced lymphocyte function. They suggest that the reduced immune responsiveness might be due to the dilution of immunocompetent cells by stem cells and immature hematopoietic cells. This suggestion is supported by their finding of increased erythropoietic tissue relative to lymphoid tissue in the spleens of *mi/mi* mice.

A chemotactic defect in peritoneal macrophages has also been demonstrated (Minkin and Pokress, 1980). In vivo, *mi/mi* mice respond to the administration of thioglycollate with only half the influx of macrophages observed in *mi/+* or *+/+* controls. In vitro, *mi/mi* macrophages respond poorly to chemotactic stimuli.

The hematopoietic and lymphoid defects caused by the *mi* mutation can be transmitted to normal mice by transplantation of affected spleen or bone marrow cells (Walker, 1975b). Mutant mice can be cured by transplanting bone marrow or spleen cells from normal littermates (Walker, 1975a,c).

Husbandry

Marks (1987) has reviewed the husbandry for this mutant. Most homozygotes die between 30 and 60 days of age, but with special care they can survive to adulthood. Soft diets are essential.

Reproduction

Only heterozygotes make effective breeders.

nu (Nude); *nu^{str}* (Streaker)

Genetics

Nude (*nu*) is an autosomal recessive mutation located on chromosome 11. The first recorded mutation at this locus occurred in a closed, but not inbred, stock of mice at the Ruchill Hospital Virus Laboratory, Glascow, Scotland (Flanagan, 1966). A second spontaneous mutation to nude occurred in the AKR/J strain at the Jackson Laboratory, Bar Harbor, Maine. This mutation was given the new name streaker and the gene symbol *nu^{str}*, which indicates both the independent occurrence of the mutation and its allelism with *nu*

A BALB/cByJ-*nu/nu* mouse. Mice homozygous for the mutation nude are athymic and hairless and vibrissae are lacking or crinkled by the time the animals are 24 hours old. Photograph courtesy of the Jackson Laboratory, Bar Harbor, Maine.

(Eicher, 1976). The nude gene has been transferred to many inbred strains. The streaker gene has also been transferred to a number of inbred strains.

Pathophysiology

The two major defects of nude homozygotes are failure of hair growth and dysgenesis of the thymic epithelium. The underlying basis for the pleiotropic effects of the nude mutation is unknown. Abnormal hair growth in nude mice has been described in detail (Eaton, 1976). Thymic dysgenesis is traceable to a developmental failure of the thymic anlage, which arises from the third pharyngeal pouch. The thymic rudiment remains small and cystic throughout life, and there are severely reduced numbers of mature functional T cells (Wortis et al., 1971). As a result, nude homozygotes do not reject allografts and often do not reject xenografts. The finding that human neoplasms can grow in nude mice has resulted in the wide use of this model for studying the mechanisms of transplantable human tumor growth and metastasis.

There is no intrinsic defect of T-cell precursors in nude mice; the T-cell defect can be corrected by transplanting mature T cells, thymocytes, or normal thymic epithelium (Wortis et al., 1971). Cytotoxic T-cell activity can be induced in nude mice by the administration of IL-2 (Hunig and Bevan, 1980), and older mice, especially if they have a microbial infection, often have some functional mature T cells. It has been suggested that the failure of stromal thymic elements to interact with lymphocytic precursors to form T cells is due to an abnormal distribution or expression of Ia antigens by epithelial components (Jenkinson et al., 1981). This is consistent with the recent finding that Ia^+ cells are absent from the thymic rudiment of nude mice (Van Vliet et al., 1985).

Nude mice respond poorly to thymus-dependent antigens because of a defect in helper T-cell activity. When responses can be detected to such antigens, antibody is largely limited to IgM. Responses to thymus-independent antigens in nude mice are normal. Levels of serum IgG1, IgG2a, IgG2b, and IgA are reduced, while IgM levels tend to be slightly elevated and IgG3 is present in normal or slightly reduced amounts. A B-cell defect has been reported, but it has not been demonstrated unequivocally (Mond et al., 1982; Wortis et al., 1982). Increased NK cell activity has also been described (Minato et al., 1980; E. A. Clark et al., 1981). Spontaneous autoimmunity has been reported in nude mice by Monier et al. (1974), who found that approximately one-third of nude homozygotes have circulating ANAs as early as 6–8 weeks of age. ANAs were present in the eluates from kidneys containing immune complexes.

Limited investigations have shown that differences in phenotypic expression between *nu* and nu^{str} homozygotes do not appear to be any greater than the reported variation among different stocks or strains of nude mice. Like nude mice, streaker homozygotes show severely reduced numbers of T cells, lack measurable T-cell activity, and show a reduction in all major classes of serum immunoglobulins (Shultz et al., 1978, 1982a). Streaker mice also have elevated NK cell activity (E. A. Clark et al., 1981). AKR/J-nu^{str}/nu^{str} mice do not show the high rate of spontaneous thymic lymphoma and chronic infection with endogenous MuLV normally present in AKR/J mice (Bedigian et al., 1979). This mutant, therefore, is a manipulatable experimental system with which to investigate the relationships among thymus function, MuLV expression, and lymphomagenesis. Although AKR-nu^{str}/nu^{str} mice do not develop T-cell lymphomas, this mutant has high numbers of preleukemic cells in the bone marrow (Shultz et al., 1983) and develops spontaneous reticulum cell sarcomas (Shultz et al., 1982a).

The mean life span of male and female streaker mice under pathogen-free conditions is 34 weeks, compared with 44 weeks for littermate controls. Moribund streaker mice show symptoms of intercurrent infections and wasting disease (Shultz et al., 1978).

Husbandry

Inbred mice homozygous for *nu* are highly susceptible to infection by a broad spectrum of bacterial and viral pathogens, and they should be maintained in a germfree, defined-flora, or pathogen-free environment (see Chapter 4). Under these conditions their life span approaches that of normal littermates. Outbred nude mice are hardier than inbred nude mice and can be maintained under less stringent conditions if isolated from conventionally housed mice. However, it should be recognized that any nude mouse housed in a conventional environment probably has a microbial infection and that this infection might influence experimental data considerably.

Streaker mice should be maintained under conditions appropriate for nude mice.

Reproduction

Neither nude nor streaker females are efficient breeders. The most effective breeding scheme uses homozygous mutant males and heterozygous females. The homozygous pups can be identified within 24 hours postpartum by their lack of vibrissae or poorly developed, crinkled vibrissae. It might be advantageous to cull some normal littermates to optimize survival of the mutant mice; however, reducing litter size below four to five pups can lead to diminished lactation. To maintain a breeding colony in conventional animal rooms, it might be desirable to graft a normal thymus under the kidney capsule of the nude males (Hetherington and Hegan, 1975).

ob (Obese)

Genetics

The mutation obese (*ob*) on chromosome 6 is a spontaneous autosomal recessive mutation discovered in 1949 at the Jackson Laboratory, Bar Harbor, Maine (Ingalls et al., 1950). The mutation is maintained on the C57BL/6J and C57BL/KsJ inbred backgrounds.

Pathophysiology

The *ob* mutation produces a marked obesity state associated with hyperphagia and hyperinsulinemia. The obesity syndrome is characterized by adipocyte hyperplasia and insulin resistance. This insulin resistance, associated with decreased numbers of high-affinity insulin receptors on liver cells, adipocytes, and lymphocytes, becomes more severe as the mice age (Kahn et al., 1973). Because manipulations, such as food restriction, that reduce hyperinsulinemia also improve insulin binding to its receptor, the reduced numbers of plasma membrane insulin receptors may be a secondary reflection of hyperinsulinemia (Soll et al., 1975).

Like the unlinked obesity gene diabetes (*db*) on chromosome 4, the pathophysiological effects of the *ob* mutation are strikingly dependent on the inbred strain background. C57BL/6J-*ob/ob* mice of both sexes develop an obesity syndrome that is well compensated by unrestricted hyperplasia of pancreatic beta cells and sustained hyperinsulinemia; males develop a mild, transient hyperglycemia between 2 and 4 months of age that is apparently corrected by the increasing pancreatic output of insulin into the serum (Hummel et al., 1972). In contrast, a severe hyperglycemia becomes established in C57BL/KsJ-*ob/ob* mice of both sexes shortly after weaning. This is associated with necrosis of pancreatic beta cells, islet atrophy, and relative insulinopenia (normal levels of serum insulin in the presence of severe insulin resistance). Unlike the transiently hyperglycemic C57BL/6J-*ob/ob* mice, which have a life span of approximately 2 years, the severely diabetic C57BL/KsJ-*ob/ob* mice usually die between 6 and 8 months of age.

Many of the perturbations in immune function in C57BL/6J-*ob/ob* mice appear to be secondary consequences of the metabolic imbalances associated with the pleiotropic mutation, because discrepancies observed in vivo between control and mutant lymphoid cell functions usually do not persist when these cells are removed from the abnormal metabolic milieu and studied in vitro. Those anomalies observed in vivo include reductions in lymphoid organ weights and in lymphocyte and monocyte numbers (Chandra and Au, 1980); depressed T-cell-mediated immunity in vivo was reflected by a reduced ability to reject allografts or to mount delayed-type hypersensitivity reactions (Sheena and Meade, 1978; Meade et al., 1979; Chandra and Au, 1980). NK cell activity and antibody-dependent cell-mediated cytotoxicity were elevated in C57BL/6J-*ob/ob* mice (Chandra and Au, 1980).

C57BL/6J-*ob/ob* mice differ from C57BL/KsJ-*db/db* mice by not exhibiting a markedly elevated PFC response after immunization with SRBCs when data are expressed on a per-cell basis. This difference very likely reflects differences in the inbred background, which include differences at the major histocompatibility complex (MHC). Both C57BL/6J-*ob/ob* and C57BL/KsJ-*db/db* mice exhibit immune complex depositions (containing antibodies against insulin) in renal glomeruli (Meade et al., 1981). Circulating autoantibodies against islet cell cytoplasmic antigens are more prevalent in C57BL/KsJ-*db/db* than in C57BL/6J-*ob/ob* mice (Yoon et al., 1988). Islet cell surface antibodies are present only at a low frequency and after initiation of the hyperglycemic phase (Flatt et al., 1985). There is no evidence that these autoantibodies are of primary pathogenic significance. However, MHC control of autoantibody expression is apparently involved because these autoantibodies are not found in a C57BL/KsJ.B6-H-2^b *db/db* congenic strain (Yoon et al., 1988).

C57BL/6J-*ob/ob* mice show an accelerated deterioration in immune function with age compared with mice with a normal genotype (Harrison et al., 1984). Presumably, compromised immune function accounts for the increased susceptibility of these mutants to diabetogenic encephalomyocarditis virus (D'Andrea et al., 1981). One contrary report claims increased immunocompetence in C57BL/6J-*ob/ob* mice, as evidenced by enhanced resistance to the B16 melanoma (Black et al., 1983).

Husbandry

Special procedures are not required to maintain these animals, although life span might be prolonged by dietary restriction.

Reproduction

Obese mice of both sexes are infertile, and breeding is done with heterozygotes. The heterozygotes are generally produced by ovarian transplantation (see Chapter 5).

oc (Osteosclerotic)

Genetics

Osteosclerotic (*oc*) is an autosomal recessive mutation located on chromosome 19 (Marks et al., 1985). It arose spontaneously in the C57BL/6J-*bf* stock at the Jackson Laboratory, Bar Harbor, Maine (Dickie, 1967).

A B6C3/Fe-*a/a oc/oc* mouse. Mice homozygous for the mutation osteosclerotic are small and have clubbed feet and a kinked tail. Photograph courtesy of the Jackson Laboratory, Bar Harbor, Maine.

Pathophysiology

Homozygous *oc* mice can be recognized at about 10 days of age. Mice are small, facial growth is stunted producing a foreshortened snout, incisors fail to erupt, feet are often clubbed, and the tail is kinked. Vestibular abnormalities cause homozygotes to circle. Homozygotes die within 30–40 days of birth.

Long bones exhibit a failure of secondary bone resorption and an absence of marrow cavities (Marks et al., 1985). Bone marrow cells are absent and leukocyte and lymphocyte counts are decreased. Erythrocyte counts, as well as the numbers of spleen and thymus cells, are within the normal range. The spleens of *oc* homozygotes are enlarged, but splenic hematopoietic stem cells are approximately normal in number, as are monoblast populations (Wiktor-Jedrzejczak et al., 1981).

Osteoclasts have poorly developed ruffled borders and lack cytoplasmic vacuolization (C. R. Marks et al., 1984). They are unable to resorb calcified cartilage (Seifert and Marks, 1985). Homozygotes exhibit rickets and have wide expanses of osteoid, unmineralized bone matrix. There is hyperplasia of the calcitonin-secreting cells of the thyroid. Serum calcium is lower than normal and does not respond to exogenous parathyroid hormone (Marks and Walker, 1969). Like the other osteopetrotic mutants, *oc/oc* mice are hypophosphatemic (Marks and Walker, 1969), have elevated serum levels of 1,25 dihydroxyvitamin D (Zerwekh et al., 1987), and have depressed levels of the 24,25 metabolite of vitamin D (Zerwekh et al., 1987).

Osteosclerotic mice are not cured by bone marrow or spleen cell transplants (Seifert and Marks, 1987). Transferability of this disease has not been studied, and little is known about immunologic dysfunction.

Husbandry

Marks (1987) has reviewed the husbandry for this mutant. Soft diets and disease-free conditions prolong the survival of homozygotes.

Reproduction

This mutation is maintained at the Jackson Laboratory in a manner identical to that of the op mutant (see pages 76–77).

op (Osteopetrosis)

Genetics

Osteopetrosis (*op*) is an autosomal recessive mutation that arose spontaneously in a C57BL/6J-*dw* stock maintained at the Jackson Laboratory, Bar Harbor, Maine (Lane, 1973). It has been mapped to chromosome 3 (Lane, 1979).

Pathophysiology

Approximately 60 percent of osteopetrotic mice that survive weaning live longer than 12 months (Marks and Lane, 1976). Homozygotes lack incisors,

A B6C3/Fe-*a/a op/op* mouse. Mice homozygous for the mutation osteopetrosis are underweight and have a dome-shaped skull.

are small, have stunted facial growth, and have a domed skull. Occasionally, hydrocephalus is found.

Like the other osteopetrotic mouse mutants, total leukocyte counts are decreased. Marrow cells are reduced to one-tenth normal. Spleen cellularity is normal, but stem cell populations are elevated (Wiktor-Jedrzejczak et al., 1981). Peritoneal macrophages and peripheral blood monocytes are markedly reduced in number (Wiktor-Jedrzejczak et al., 1982).

Bone in *op* homozygotes contains large lipoid droplets of unknown origin. Osteoclasts are small and markedly reduced in number, but they have elaborate ruffled borders and extensive cytoplasmic vacuolization (Marks, 1982). They exhibit an abnormal acid phosphatase distribution within the cell and have lipoid inclusions that might be related to the larger extracellular lipoid masses (Marks and Lane, 1976). Bone matrix formation and parafollicular (calcitonin-secreting) cells of the thyroid are increased, serum phosphate levels are decreased, and serum calcium levels are normal (Marks and Walker, 1969).

Olsen et al. (1978) have reported that spleen cells from *op* homozygotes have defective in vitro responses to PHA, LPS, and poly(I-C); however, it has been suggested more recently (Schneider and Marks, 1983) that the reduced immune responsiveness might be due to the dilution of immunocompetent cells by stem cells and immature hematopoietic cells (see also *mi*, page 67).

The disease induced by the *op* mutation undergoes spontaneous partial resolution in adulthood; however, these mice cannot be cured by transplantation of normal bone marrow or spleen cells (Marks et al., 1984). Conversely, the transfer of spleen cells from *op* homozygotes to lethally irradiated, normal littermates does not transfer the defect (S. C. Marks, Jr., University of Massachusetts Medical School, unpublished data). Normal mouse fibroblast culture medium has been reported to stimulate spleen cells from *op/op* mice to differentiate into monocytes and macrophages, which suggests that the op defect might be an abnormality of the microenvironment (Wiktor-Jedrzejczak et al., 1982).

Husbandry

Marks (1987) has reviewed the husbandry for this mutant. Homozygous *op* mice survive weaning if they are provided soft food; however, viability is reduced, despite the partial resolution of the defect in adulthood. Specific-pathogen-free (SPF) conditions are recommended.

Reproduction

Homozygotes do not breed; breeding can be accomplished by using heterozygotes. This mutation is propagated at the Jackson Laboratory by a system of ovarian transplantation using compatible sets of hybrids (see Chapter 5).

scid (Severe Combined Immunodeficiency)

Genetics

Severe combined immunodeficiency (*scid*) is an autosomal recessive mutation that occurred spontaneously in the C.B-*Igh-1^b* (N17F34) (usually abbreviated CB-17) congenic strain (Bosma et al., 1983). In this congenic strain the inbred BALB/c strain carries the immunoglobulin heavy-chain allele (*Igh-1^b*) of the C57BL/Ka strain. The *scid* locus has recently been mapped to the centromeric end of chromosome 16 (Bosma et al., 1989).

Pathophysiology

Homozygous *scid* mice have little or no immunoglobulin in their serum. The lymph nodes and thymus are abnormally small, as is the spleen of most animals. The thymus consists of a rudimentary medulla without a cortex. Spleen and lymph node follicles are virtually devoid of lymphocytes. All of these lymphoid organs consist primarily of vascularized supportive tissue with variable numbers of fibroblasts, histiocytes, and macrophages. Bone marrow, although lacking lymphocytes and plasma cells, appears otherwise to be morphologically normal (Bosma et al., 1983; Custer et al., 1985).

Mice with the *scid* mutation lack dendritic Thy-1$^+$ epidermal cells (Nixon-Fulton et al., 1987). They cannot reject allogeneic grafts or produce an-

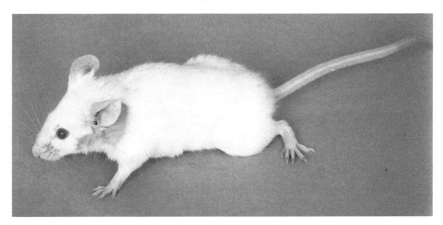

A C.B-*Igh-1^b*/*Igh-1^b scid/scid* mouse. Mice homozygous for the mutation severe combined immunodeficiency have no abnormal external characteristics when they are maintained under germfree conditions. Photograph courtesy of the Jackson Laboratory, Bar Harbor, Maine.

tibodies to common laboratory antigens, and their spleen cells do not proliferate in response to T- or B-cell-specific mitogens. Fluorochrome-conjugated antibody reagents that specifically stain B and pre-B cells fail to detect such cells in the spleen or bone marrow (Bosma et al., 1983; Dorshkind et al., 1984). The presence of early B cells is, nonetheless, evident from the ability of Abelson murine leukemia virus (A-MuLV) to transform *scid* bone marrow cells (Fulop et al., 1988). The resulting A-MuLV transformants show a pre-B-cell phenotype, that is, a rearrangement of both *Igh* alleles with the *Igl* alleles remaining in the germline configuration (Schuler et al., 1986).

There is also evidence for the presence of early T cells, as about 15 percent of *scid* mice spontaneously develop thymic lymphomas (Custer et al., 1985) that contain rearranged *Tcr*γ and *Tcr*β alleles and express T-cell-specific cell surface antigens (L3T4, Ly-2) (Custer et al., 1985; Schuler et al., 1986). However, the observed *Igh* and *Tcr* rearrangements in the thymic lymphomas and A-MuLV transformants, respectively, are abnormal, consisting of aberrant deletions of variable-region-coding elements (V, D, or J). The deletions appear to result from the attempted rearrangement of these elements, suggesting that the VDJ recombinase mechanism is defective (Schuler et al., 1986). This idea is further supported by several recent findings. First, unrearranged *Igh*, *Igl*, and *Tcr* genes become transcriptionally active in lymphopoietic tissues, but fail to produce functional transcripts (Schuler et al., 1988). Second, rearranged *Igh* alleles in A-MuLV transformants lack D elements, J elements, or both at the recombination sites (Hendrickson et al., 1988; Kim et al., 1988; Malynn et al., 1988). Similar observations have been made in B-cell lines of long-term *scid* bone marrow cultures (Okazaki et al., 1988). Finally, transformed pre-B and pre-T cells contain an active but abnormal VDJ recombinase activity that is unable to catalyze the formation of functional joints between V(D)J elements (Lieber et al., 1988).

As an inbred mutant strain, all *scid* mice share the same single genetic disease, and yet not all mice lack functional lymphocytes. A variable percentage of young adult mice (2–20 percent) appear "leaky" in that they develop low numbers of functional B and T cells (Bosma et al., 1988; Carroll and Bosma, in press). This condition is not inherited and may reflect a very low rate of somatic reversions at either *scid* allele, such that the VDJ recombinase activity is normalized in some developing lymphocytes in *scid* homozygotes. Alternatively, the leaky phenomenon may reflect a low but finite chance of a good rearrangement at two critical antigen receptor loci (e.g., *Igh* and *Igl*) in cells with a highly defective VDJ recombinase system.

The *scid* mutation does not appear to affect the differentiation of myeloid cells (e.g., erythrocytes, granulocytes, and macrophages). The capacity of *scid* splenic cells to generate mixed myeloid colonies in vitro is equivalent to that of normal C.B-*Igh-1*^{b} mice (Dorshkind et al., 1984). Moreover, macrophage activation and antigen-presenting function are unimpaired (Czitrom et al., 1985; Bancroft et al., 1986). Interestingly, NK cell activity is also unaffected by the *scid* mutation (Dorshkind et al., 1985; Hackett et al., 1986). These NK cells are capable of mediating the rejection of allogeneic bone marrow grafts transplanted into irradiated *scid* hosts (Murphy et al., 1987).

Homozygotes can be "cured" of their lymphocyte deficiency by engraftment with histocompatible bone marrow or fetal liver cells of normal mice (Bosma et al., 1983). For example, injection of neonatal *scid* mice with $\geq 0.5 \times 10^6$ fetal liver cells results in full lymphoid reconstitution at 8–12 weeks of age (R. P. Custer, G. C. Bosma, and M. J. Bosma, Institute for Cancer Research, Fox Chase, Pa., unpublished data). Spleen and lymph node follicles become repopulated with lymphocytes, and Peyer's patches and solitary follicles of the intestinal tract return to normal. The thymus shows a prominent cortex and corticomedullary delineation, and there is a normal morphologic gradient of developing lymphocytes that proceeds from capsule to medulla. Injection of adult *scid* mice with 3×10^6 to 5×10^6 normal bone marrow cells also results in lymphocyte repopulation, although the reconstitution is often incomplete (Custer et al., 1985). This problem can be circumvented by sublethal irradiation of adult mice prior to cell transfer (Fulop and Phillips, 1986). Recently, the *scid* mouse has been used as a recipient for transplanted human tumors (Reddy et al., 1987), human fetal lymphoid tissues (McCune et al., 1988), and peripheral blood lymphocytes (Mosier et al., 1988). Thus, the *scid* mouse represents a potential new model for engraftment of xenogeneic cells and tissues.

The *scid* mouse is also useful for examining the relationship between immunity and disease. Specifically, it serves as a model for studying human severe combined immune deficiency and its associated infections. As in humans, *scid* mice develop a severe interstitial pneumonitis resulting from infection by *Pneumocystis carinii*. This microorganism is easily demonstrated with Gomori's methenamine silver stain (J. B. Roths and C. L. Sidman, The Jackson Laboratory, Bar Harbor, Maine, unpublished data). In addition, the high incidence of spontaneous thymic lymphomas makes the *scid* mouse a potential model for understanding the basis of increased lymphoid malignancies in certain immune deficiency states.

Husbandry

Homozygous *scid* mice readily succumb to microbial infections because of their lack of an immune system and must be maintained in a pathogen-free environment (see Chapter 4). Those animals maintained in a single-user, SPF, barrier-protected room can survive to 9–12 months of age.

Reproduction

Homozygous *scid* mice can be bred without difficulty, although the average litter size (four to six) is smaller than that of the congeneic C.B-*Igh-1^b* strain (six to nine).

Tol-1 (Tolerance to Gamma-Globulin)

Genetics

The *Tol-1* locus controls the induction of tolerance to bovine gamma-globulin (BGG) or human gamma-globulin (HGG). Induction of tolerance to BGG is easy in the DBA/2 strain and difficult in the BALB/c strain (Das and Leskowitz, 1970). F1 hybrid and backcross analyses suggest that the difference is due to alleles at a single autosomal locus (Lukić et al., 1975).

Pathophysiology

Cell transfer studies indicate that the difference in induction of tolerance to gamma-globulin is a function of macrophages (Das and Leskowitz, 1974). In vitro studies show that macrophages of the BALB/c mouse can efficiently remove a minor component of BGG, which renders the BGG capable of easily inducing tolerance (Cowing et al., 1977). Preliminary studies with HGG suggest that IgG3 is the critical component (Cowing et al., 1979).

Husbandry

Special husbandry procedures are not required.

Reproduction

These animals reproduce normally.

vit (Vitiligo)

Genetics

Vitiligo (*vit*) is an autosomal recessive mutation that was first identified by E. S. Russell at the Jackson Laboratory, Bar Harbor, Maine, as a spontaneous mutation that occurred in the C57BL/6J strain. It is now maintained as C57BL/6JLer-*vit*/*vit* (Palkowski et al., 1987).

Pathophysiology

Mice homozygous for the *vit* locus have congenital white spotting of the back and abdomen and a progressive depigmentation with each spontaneous hair molt. By 2 years of age the mutant mouse is white. The epidermis, including that of the tail and ear, loses all pigment with age (M. Bell et al., 1984). Light and electron microscopic changes include degeneration of follicular melanocytes without inflammation. There is also destruction of melanocytes with the choroid and pigment epithelium of the eyes (Palkowski et al., 1987).

C57BL/6JLer-*vit*/*vit* mice have fewer Ia-positive Langerhans' cells in the interfollicular epidermis of the back and ear than do C57BL/6J control mice (Rheins et al., 1986). However, the number of Langerhans' cells present in the follicullar epidermis is normal and unchanged during depigmentation (Palkowski et al., 1987). The density of epidermal Thy-1^+ cells increases during depigmentation (Amornsiripanitch et al., 1988). While young pigmented C57BL/6JLer-*vit*/*vit* mice were capable of developing moderate contact sensitivity (allergic contact dermatitis) to 2,4-dinitro-1-fluorobenzene (DNFB) and picryl chloride, older depigmented animals were not (Rheins et al., 1986). Transplantion of normal C57BL/6 mouse skin onto a depigmented C57BL/6JLer-*vit*/*vit* mouse restores the ability to induce contact sensitivity (Palkowski et al., 1987), and transplantation of *vit*/*vit* skin onto normal C57BL/6 mice fails to allow the induction of contact sensitivity, suggesting that the contact sensitivity defect in depigmented mice is cutaneous, probably epidermal, and not systemic in origin (Amornsiripanitch et al., 1988). Since the contact sensitivity reaction is a form of cell-mediated immunity, the delayed-type hypersensitivity reaction was evaluated in *vit*/*vit* mice.

Both C57BL/6JLer-*vit*/*vit* and normal C57BL/6J mice were sensitized with SRBCs given intravenously and challenged with intradermal SRBC injections. Young *vit*/*vit* mice had delayed hypersensitivity reactions comparable to those of age-matched controls; however, 24-week-old depigmented *vit*/*vit* mice demonstrated enhanced reponsiveness (Amornsiripanitch et al., 1988). The C57BL/6JLer-*vit*/*vit* mouse was also comparable to C57BL/6J in the production of humoral immunity to T-cell-dependent and -independent antigens. Skin graft rejection also appears to be normal in *vit*/*vit* mice (Amornsiripanitch et al., 1988).

When pigmented skin from a C57BL/6J mouse is transplanted onto a depigmenting *vit*/*vit* mouse, it does not depigment, suggesting that the process of melanocyte degeneration is not systemic (Lerner et al., 1986).

Husbandry

Information has not been published.

Reproduction

Information has not been published.

W (Dominant Spotting); W^v (Viable Dominant Spotting)

Genetics

The W locus is found on chromosome 5. Associated with this locus are many mutant alleles, most of which, including W and W^v, are semidominant in expression. The W mutation occurred many years ago and was preserved by mouse fanciers (Dunn, 1937). The W^v mutation was discovered in the C57BL strain in 1937 (Little and Cloudman, 1937). Dominant spotting is maintained as a heterozygote (W/+) on the C57BL/6J and WB/ReJ backgrounds. Viable dominant spotting is maintained on the C57BL/6J, MWT/Le, NFR/N, NFS/N, and WB/ReN backgrounds. The compound mutant (W/W^v) is maintained on the WB/ReJ × C57BL/6J F_1 (abbreviated WBB6F_1) hybrid background.

Pathophysiology

W homozygotes or W/W^v compound mutant mice are white with black eyes, are sterile, and have macrocytic anemia. Although W homozygotes have severe anemia from day 12 of gestation and die within the first week

A mouse carrying the mutation viable dominant spotting (W^v). Affected mice have white spotting and a slightly dilute coat color. Photograph courtesy of the Jackson Laboratory, Bar Harbor, Maine.

of birth (Russell, 1970), W^v homozygotes can survive to maturity. Heterozygotes ($W/+$ or $W^v/+$) have white spotting and a slightly diluted coat color, have normal erythrocyte numbers ($W/+$) or a slight macrocytic anemia ($W^v/+$), and are fully viable and fertile (Russell, 1949).

Macrocytic anemia and pigment defects are due to intrinsic defects in progenitor cells of erythrocytes and melanocytes. Some evidence suggests that the defect operates at a stage before the proliferation of melanoblasts in the skin (Mayer, 1970). The defect is also present in $W/+$ heterozygotes (Gordon, 1977). The genetic basis for anemia probably resides in erythropoietic cells at an erythropoietin-sensitive stage (Bannerman et al., 1973) and appears to involve a Thy-1$^+$ regulatory cell that is necessary for stem cell differentiation (Sharkis et al., 1978). Lymphopoietic cells also appear to be abnormal. A lymphocytic stem cell population absent in W^v homozygotes can be demonstrated in lymph nodes of homozygotes transplanted with normal bone marrow (Harrison and Astle, 1976). The W/W^v compound mutant has a severe reduction in the number of mast cells (Kitamura et al., 1978). Aplastic anemia in W homozygotes is corrected by transplantation of bone marrow, spleen, or fetal liver stem cells from syngeneic ($W/+$ or $+/+$) donors. Irradiation is not required prior to transplantation. Erythrocyte, as well as granuloctye, populations become the donor type following transplantation (Murphy et al., 1973).

Recently a new mutation, W^f, has occurred at this locus. Homozygotes have normal viability and fertility and no anemia. These mice, however, still lack mast cells (Stevens and Loutit, 1982).

W/W^v mice have a delayed expulsion of primary nematode (*Nippostrongylus brasiliensis*) infestation but are refractory to secondary infestation (Crowle and Reed, 1981). When reconstituted with bone marrow or spleen cells from normal littermates, they develop normal connective tissue mast cells. They also acquire mucosal mast cells in response to parasitic infection, but the mucosal mast cells do not accelerate parasite rejection (Crowle, 1983). Likewise, W/W^v mice have a delayed expulsion of *Trichinella spiralis* (Ha et al., 1983); however, in this instance bone marrow transplantation, which reconstitutes the mast cell population, does accelerate parasite rejection.

Husbandry

Special husbandry procedures are not required for maintaining W/W^v compound mutants or W^v homozygotes. Special husbandry procedures will not prolong the life of W homozygotes.

Reproduction

The mutation dominant spotting is maintained by breeding heterozygotes. W/W^v mice are produced by mating WB/ReJ-$W/+$ females with C57BL/6J-

$W^v/+$ males. These mice are easily distinguished from $W^v/+$, $W/+$, and $+/+$ littermates by the distinctive white coat color (Crowle and Reed, 1981). $W^v/_s$ and $W/+$ heterozygotes can be distinguished from the wild type ($+/+$) by the presence of a white spot on the abdomen.

wst (Wasted)

Genetics

Wasted (*wst*) is an autosomal recessive, lethal mutation that arose spontaneously in 1972 in the HRS/J strain at the Jackson Laboratory, Bar Harbor, Maine (Lane, 1981). It has been mapped to chromosome 2 (Sweet, 1984). The *wst* mutation was crossed to a segregating F1 hybrid background (C3HeB/FeLe-*a/a* × C57BL/6J F_1) to increase the viability of progeny (Shultz et al., 1982b). The gene is being transferred to strain C57BL/6J (N11 as of December 1988) (L. D. Shultz, The Jackson Laboratory, Bar Harbor, Maine, unpublished data).

Pathophysiology

Homozygous *wst* mice are recognized by 20 days of age by neurological abnormalities, including tremors and ataxia. Paralysis precedes death at 30 days of age or less. Wasted mice gain weight only until 20 days of age. By 28 days, *wst* homozygotes are one-half as large as $+/-$ controls.

Homozygous *wst* mice show pathologic changes similar to those of patients with ataxia telangiectasia. Central nervous system abnormalities include degeneration of cerebellar Purkinje cells and focal demyelination in the cerebellar cortex and ventral columns of the spinal cord. There is a fourfold greater incidence of chromosome damage in bone marrow cells of *wst* homozygotes than in those of heterozygous littermates.

Both thymus-dependent and thymus-independent areas of the lymph nodes are markedly hypoplastic. The follicles of the lymph nodes and the spleen are poorly developed, Peyer's patches are reduced in size, and the thymic cortex has reduced cellularity. In addition, *wst/wst* mice have a threefold decrease in numbers of circulating leukocytes. Analysis of lymphocyte subsets by immunofluorescence microscopy indicates that the percentages of B and T cells are within the normal range.

Homozygous *wst* mice have significantly impaired delayed-type hypersensitivity to SRBCs when tested 4 days following priming (Shultz et al., 1982b). The response of thymic and splenic lymphocytes to ConA or LPS is markedly reduced (Goldowitz et al., 1985). Wasted mice have normal levels of serum IgA, and the spleen has normal numbers of sIgA-bearing B cells and plasma cells. However, IgA plasma cells are totally absent from the intestine, and IgA-specific B-cell precursors are absent from the Peyer's patches. Kaiserlian et al. (1985) suggest that the *wst* mutation might be useful as a model for the regulation of IgA production.

Nordeen et al. (1984) have expressed reservations about wasted mice being considered as a model for human ataxia telangiectasia. The gamma irradiation- or bleomycin-resistant component of DNA replication characteristic of fibroblasts in humans with ataxia telangiectasia (Cramer and Painter, 1981) has not been observed in fibroblast cell lines of wasted mice (Nordeen et al., 1984).

Recently, Abbott et al. (1986) described abnormally low levels of adenosine deaminase (ADA) in wasted mice and proposed that *wst* represents a mutation in the structural gene for this enzyme. However, investigations by Geiger and Nagy (1986) indicate that these mice have measurable ADA levels in blood and lymphoid tissues and increased ADA activity in the spleen and cerebellum. Thus, the status of the wasted mouse as an animal model for human ADA deficiency is still in question.

Husbandry

Special husbandry procedures are not required.

Reproduction

Wasted mice can be perpetuated by mating heterozygotes. At the Jackson Laboratory the mutation is maintained by ovarian transplantation (see Chapter 5).

xid (X-Linked Immune Deficiency)

Genetics

X-linked immune deficiency (*xid*) is a sex-linked recessive mutation that originated on the CBA/HN substrain (Amsbaugh et al., 1972; Huber et al., 1977). This mutation is now available on several genetic backgrounds.

Pathophysiology

The mutation *xid* causes functional defects of B lymphocytes. Homozygous females and hemizygous males (hereafter called *xid* mice) fail to respond to thymus-independent antigens such as haptenated-Ficoll, dextran, pneumococcal polysaccharide, and dsDNA (Scher et al., 1973; Cohen et al., 1976). Furthermore, they are nonresponsive to certain thymus-dependent antigens (Press, 1981). Affected mice have low levels of the immunoglobulin isotypes μ and γ_3 and extremely low levels of λ. The B-cell defect is believed to be intrinsic, because the functional lesions are overcome when normal bone marrow is transferred to an *xid* host. Conversely, *xid* marrow transplanted to an irradiated, normal, syngeneic host retains its abnormal phenotype.

Affected mice lack a subpopulation of B cells that expresses small amounts of surface IgM and large amounts of surface IgD. This subpopulation also expresses a unique surface component (Lyb-3) that, when bound by antibody, induces an enhanced response to antigen, which suggests that it functions as a lymphokine receptor (Huber, 1982). The same subset of B cells expresses the alloantigen Lyb-5 (Smith et al., 1986). In normal mice this B-cell subset responds to thymus-independent antigens, is polyclonally stimulated by anti-immunoglobulin, forms colonies in vitro (Kincade, 1977), binds to macrophage surfaces, and is exquisitely radiation sensitive (Lee and Woodland, 1985). It is a matter of controversy whether these B cells form a unique differentiation pathway or represent a late maturation stage of all normal B cells (see *nu xid*, page 92). It has also been argued that those B cells that are present in *xid* mice are not normal (Sprent and Bruce, 1984). The Ly-1^+ B-cell population is also missing in these mice (Hayakawa et al., 1983).

Mice of the H-2^b haplotype carrying the *xid* mutation fail to express the surface antigen IaW39 but do have IaW39-positive cytoplasmic protein. This suggests that the failure to express IaW39 as a surface molecule is due to a deficiency in a glycosylation pathway (Huber, 1982). By restriction fragment length polymorphism, a family of X-linked genes located close to or at the *xid* locus has been mapped. It appears that mature normal B cells, but not *xid* B cells, transcribe gene products that are encoded in this region (Cohen et al., 1985).

No defects in T-cell functions, such as cytotoxicity, graft rejection, or delayed-type hypersensitivity reactions, have been observed, but there are several studies suggesting that the help provided by T cells from *xid* mice is not optimal (cf. Phillips and Campbell, 1981). Agreement on this point has not been achieved.

When the *xid* allele is backcrossed onto a C3H/HeN or a C3H/HeJ background, the affected mice have a greater defect in B-cell function than do CBA/HN-*xid* mice. Responses to the mitogen *Nocardia* are diminished, and the frequency of sIg$^+$ B cells, particularly in the lymph nodes, is greatly reduced. It has been proposed that those B cells present are nonresponsive to some of the T-cell-generated lymphokines. The major functional defect so far described (in addition to those already ascribed to *xid*) is a diminished B-cell response to type I antigens (Mond et al., 1983).

Backcrossing the *xid* allele onto the NZB (Nakajima et al., 1979), NZB × NZW F_1 (A. D. Steinberg et al., 1984), or BXSB (A. D. Steinberg et al., 1984) backgrounds results in a modification of the natural history of autoimmune disease that is characteristic of these strains. All have reduced quantities of autoantibodies and increased life spans.

Husbandry

Special husbandry procedures are not required.

Reproduction

These animals breed normally and can be maintained by brother × sister matings.

Yaa (Y-Linked Autoimmune Accelerator)

Genetics

The Y chromosome linkage for acceleration of the autoimmune disease of male mice of the SB/Le and BXSB/Mp strains was first hypothesized from studies of reciprocal F1 hybrid males resulting from crosses to strains C57BL/6J, NZB/BlNJ, and SJL/J (Murphy and Roths, 1979). Conclusive evidence for this holandric inheritance of accelerated autoimmunity came from analysis of the Y-consomic stocks developed by transfer (backcross) of the Y chromosome of strain BXSB to strains C57BL/6J and NZB/BlNJ and the transfer of the normal Y chromosome of strain C57BL/6J to strain BXSB/Mp. The gene symbol *Yaa* (Y-linked autoimmune accelerator) has been assigned (Roths, 1987).

Yaa acts synergistically with a presumptive autosomal autoimmune induction gene(s) in strains SB/Le and BXSB/Mp. The development and genetics of these strains are described in the section on immunodeficient inbred strains (see page 96).

Pathophysiology

BXSB-*Yaa* males are generally 10–15 percent smaller than their female siblings prior to and at weaning. They develop moderate lymphadenopathy, which is most noticeable as enlargement of the cervical and axillary lymph nodes, by 3–4 months of age. The spleens are greatly enlarged and easily palpable. Greater than 60 percent of BXSB/Mp-*Yaa* males develop a profound generalized edema within a few days prior to death. BXSB/Mp-*Yaa* males survive to 4–6 months compared with 14–16 months for female BXSB siblings. They remain healthy until near the end of their short life span without evidence of infectious disease. Male BXSB mice bearing the normal Y chromosome of C57BL/6J (i.e., BXSB/Mp-+^{Yaa}) live for greater than 24 months without evidence of autoimmune disease.

Yaa operates independently of the sex hormonal milieu. For example, testes weights, serum testosterone levels, and secondary sex characteristics are not altered by *Yaa* (Roths, 1987). In addition, Eisenberg and Dixon (1980) have demonstrated that orchiectomy does not delay the early onset of fatal autoimmune disease of BXSB males.

The moderate lymph node enlargement is due to proliferating lymphocytes and admixed plasma cells and histiocytes. Lymph node architecture is blurred. Immunoblasts are common and microvascular proliferation may occur. The enlarged spleens show moderate hyperplasia of white pulp and greatly increased erythropoiesis. The kidneys of BXSB/Mp-*Yaa* males show an acute or subacute exudative and proliferative glomerulonephritis and striking tubular involvement. Immune complex vascular disease is also common. Immunofluorescence studies of kidneys of BXSB males reveal a heavy granular deposition of immunoglobulins in the glomerular capillary walls and mesangium.

Nearly all BXSB/Mp-*Yaa* males are clinically anemic (packed erythrocyte volumes of 30–33 percent at 6 months of age), with high titers of circulating erythrocyte autoantibody and elevated protoporphyrin. High titers of antinuclear and thymocyte-binding autoantibodies are also characteristic. Hyperimmunoglobulinemia (17 percent of total globulins) with ninefold elevations of the IgA and IgG1 isotypes has been described (Murphy and Roths, 1978a).

There is an increase in the frequency and absolute numbers of immunoglobulin-bearing (B-cell) lymphocytes in the lymph nodes by 4 months of age (Theofilopoulos et al., 1979). The *Yaa* gene appears to play a significant role in the abnormal resistance to tolerance induction to heterologous IgG (Izui and Masuda, 1984). Cell-mediated immune functions of BXSB/Mp-*Yaa* males are similar to those of nonautoimmune control strains, but BXSB males have reduced reticuloendothelial function and lower bactericidal activity against *Listeria* sp. (Creighton et al., 1979).

The *Yaa* mutation has been transferred by multiple backcross matings to C3H/HeJ, C57BL/6J, and SJL/J inbred strains. Compared with their consomic controls, C3H/HeJ-*Yaa* and SJL/J-*Yaa* males do not develop autoimmune manifestations. C57BL/6J-*Yaa* males, but not C57BL/6J-+^{Yaa} males, develop antinuclear autoantibodies, have a fourfold increase of serum IgG2b, and have splenomegaly (58 percent increase in mass). The mean life span of C57BL/6J-*Yaa* males is 80 percent that of their consomic controls (23 versus 29 months) (Roths, 1987). The *Yaa* gene is capable of accelerating the appearance of the anti-DNA response induced by the *lpr* gene (Pisetsky et al., 1985).

Husbandry

Special husbandry procedures are not required.

Reproduction

BXSB/Mp females give birth to an average of five pups per litter, and in paired matings they will produce only two to three litters because of the early morbidity of males.

S-Region-Linked Genes Controlling Murine C4

Genetics and Pathophysiology

Mice have two C4-like molecules, each of which is encoded by a separate gene in the S region of the MHC (Shreffler, 1982). One glycoprotein, coded for by the sex-limited protein (*Slp*) locus, is structurally similar to human and guinea pig C4, is testosterone dependent in many strains, but has no hemolytic activity (Hobart, 1984). A second closely linked gene, *C4* (formerly designated *Ss*), encodes for the antigenic, structural, and functional homologue of human C4 (Ferreira et al., 1977). Mice expressing the H-2^{w7} haplotype have C4 that has only 30 percent of the functional complement activity of other strains (Atkinson et al., 1980). The alpha chain of this C4 is smaller in H-2^{w7} mice because of a difference in the carbohydrate content (Karp et al., 1982a,b). This genetic variation in glycosylation of C4 has been shown to directly affect hemolytic activity, because evidence for a plasma inhibitor or altered C4 stability is lacking (Atkinson et al., 1980). Recent evidence indicates that mice with the H-2^{w7} haplotype have four *Slp* genes, three of which appear to be C4-*Slp* recombinant genes (Nakayama et al., 1987). Synthesis of the products of these genes is not testosterone dependent, and at least two *Slp* genes of H-2^{w7} mice are transcribed (Ogata and Sepich, 1985). The relationship between products of these recombinant genes and the H-2^{w7}-linked C4 gene product with faulty glycosylation is unclear.

A quantitative difference in C4 is also controlled by a single autosomal gene with codominant expression. The original locus designation was *Ss* and the alleles were *h* (high levels) and *l* (low levels), that is, Ss^h and Ss^l. Current nomenclature for the *Ss* locus is *C4*; however, the alleles described above have not appeared associated with the new nomenclature in the literature. Ss^l strains are all of the H-2^k haplotype (Shreffler and Owen, 1963) and have 20-fold less functional C4 in the circulation. F1 hybrids have intermediate levels. No functional impairment in complement activation or resistance to infectious disease has been attributed to the Ss^l allele alone in mice that bear it.

Another locus or gene cluster linked to the S region of mice controls the quantitative levels of complement components 1, 2, and 4. The data suggest that functional levels of at least these three components are controlled by a single gene or a small gene cluster within the *H-2* complex (Goldman and Goldman, 1976). In humans, a gene or closely linked gene cluster controls expression of factor H, C4-binding protein, and the C3b receptor. This gene is called the regulator of complement activation (RCA), and perhaps similar loci regulate the expression of multiple complement genes in mice (De Cordoba and Rubinstein, 1986).

Husbandry

Special husbandry procedures are not required.

Reproduction

No reproductive problems have been attributed to the various loci controlling the expression of the early components of complement described above.

MICE WITH MULTIPLE MUTATIONS[1]

gld (Generalized Lymphoproliferative Disease); *xid* (X-Linked Immune Deficiency)

C3H/HeJ-*gld/gld xid* mice have been developed to determine the effect of the *xid* gene on the development of the *gld* autoimmune lymphoproliferative phenotype (Seldin et al., 1987). A comparison of C3H-*gld/gld* and C3H-*gld/gld xid* males shows that the *xid* gene has no effect on the extent of lymphadenopathy or on the phenotype of the expanded T-cell subset (Ly-2$^-$, Ly-1$^+$, L3T4$^-$, Ly-5$^+$) of *gld* homozygotes. By contrast, the *xid* gene significantly reduces serum IgM and nearly abolishes the generation of both IgM and IgG anti-ssDNA and anti-dsDNA autoantibodies. Thus, the *xid* gene dramatically decreases B-cell manifestations without affecting T-cell abnormalities in *gld* homozygotes.

lpr (Lymphoproliferation); *nu* (Nude)

The C57BL/6J-*lpr/lpr nu/nu* double mutant strain has recently been developed. The *lpr* phenotype on the C57BL/6J background has been defined

---

[1]The husbandry and reproduction of any double mutation carrying *nu/nu* are like those discussed in the narrative on nude mice (pgs. 69–71 and Chapter 4).

(Mosbach-Ozmen et al., 1985a). Homozygosity at the *nu* locus prevents the development of *lpr*-induced lymphadenopathy (a T-cell hyperplasia) (Mosbach-Ozmen et al., 1985b). This finding is consistent with the inhibition of lymphadenopathy seen in neonatally thymectomized MRL/Mp-*lpr/lpr* mice (A. D. Steinberg et al., 1980).

lpr (Lymphoproliferation); *xid* (X-Linked Immune Deficiency)

MRL/Mp-*lpr/lpr xid* mice have been developed (E. B. Steinberg et al., 1983). Double mutant *lpr/lpr xid* mice show the lymphadenopathy, expansion of a dull Ly-1$^+$ T-cell population, impaired cellular proliferation in response to ConA, and diminished IL-2 production characteristic of homozygous *lpr* mice. However, the double mutants do have a marked reduction in anti-ssDNA and anti-dsDNA autoantibody levels and serum gp70 immune complexes. At 6 months of age, histologically defined renal disease and proteinuria are markedly reduced in *lpr/lpr xid* mice, and their life span is nearly double that of mice homozygous for the *lpr* gene alone. Possibly, this is due to the effect of the *xid* mutation on the maturation of the B-cell subset necessary for high autoantibody production, which requires nonspecific T-cell helper factors (Fieser et al., 1984).

lpr (Lymphoproliferation); *Yaa* (Y-Linked Autoimmune Accelerator)

The Y-linked autoimmune accelerator (*Yaa*) gene, originally identified in studies of strain BXSB, is capable of promoting or accelerating autoimmune disease induced by genes other than those resident in the BXSB autosomal genome. In C57BL/6J-*lpr/lpr Yaa* double mutant males, *lpr*-induced lymphadenopathy occurs earlier and is more extensive than in homozygous *lpr* controls. Moreover, the survival time of *lpr/lpr Yaa* mice is reduced to 9 months, compared with 15 months for males homozygous for *lpr* alone (Roths, 1987). C57BL/6J-*lpr/lpr Yaa* males have markedly increased levels of IgG anti-DNA (8.2 $\log_2$ titers) in comparison with *lpr* homozygotes without *Yaa* (2.5 $\log_2$ titers). Serum IgM and IgG levels are 1.5- and 2.1-fold greater, respectively, in males with both mutations than in males homozygous for *lpr* alone (Pisetsky et al., 1985).

nu (Nude); *bg* (Beige)

Comparisons between homozygous N:NIHS-*nu/nu* and double homozygous N:NIHS-*bg/bg nu/nu* mice reveal that NK cell activity in double homozygotes is significantly reduced compared with that in nude homozygotes and three times greater than that in beige homozygotes (Fodstad et al., 1984a).

An evaluation of the double homozygotes as hosts for human tumor transplantation has shown no correlation between host NK cell activity and subcutaneous growth of various human (LOX, CEM, and K562) and murine (YAC-1) tumor cells (Fodstad et al., 1984b). In addition, low NK cell activity has not been associated with increased lung colony formation in a metastasis model in which intravenously injected human LOX or murine B16F10 melanoma cells are used. Thus, the data failed to support the idea that NK cells exert significant effects on tumor cells in vivo, although they are known to be toxic to tumor cells in vitro.

nu (Nude); *xid* (X-Linked Immune Deficiency)

Mice with these mutations have no mature T cells and few or no mature B cells (Azar et al., 1980). The result is a defect resembling severe combined immunodeficiency (Mond et al., 1982; Wortis et al., 1982). B-cell precursors are blocked at the pre-B-cell stage, at which point they express the cell surface antigen Lyb-5, have incompletely rearranged immunoglobulin genes, and are insensitive to transformation by A-MuLV (Karagogeos et al., 1986). Grafts of thymus epithelium restore both T- and B-cell function, demonstrating that B cells expressing the *xid* mutation depend on T cells or their products to mature past the pre-B-cell stage (Karagogeos and Wortis, 1987). There are two explanations for this synergistic effect. One is that all B cells in *xid* mice are "crippled" and need the extra stimulation provided by T cells to reach maturity. The alternative is that normal B cells develop into two separate subpopulations, one of which (found in normal and *xid* mice) is T-cell dependent and the other (found in normal and *nu/nu* mice) is T-cell independent.

Histologic examination of 173 N:NIHS-*nu/nu xid/xid* female mice revealed a high incidence of both lymphosarcoma and ovarian granulosa cell tumors (Sadoff et al., 1988).

nu (Nude); *bg* (Beige); *xid* (X-Linked Immune Deficiency)

The simple mode of inheritance of these mutations provides a means of designing experimental systems that allow the study of the effects of these mutants either singly or in various combinations in the same experiment. The technique involves developing congenic strains of each mutant on the same genetic background. First, two of these mutations and finally all three are combined. The combination used depends on the purpose of the study. The immune system effects of the possible combinations are given in Table 2-2.

This scheme meets all the criteria for good experimental design and is also flexible, because selected combinations can be used without compromising

TABLE 2-2 Cells Present in Mutant Combinations of *bg*, *nu*, and *xid*

Mutant Locus			Cells Present		
bg	nu	xid	T	B	NK
+/+	+/+	+/+	+	+	+
bg/bg	+/+	+/+	+	+	−
+/+	nu/nu	+/+	−	+	+
+/+	+/+	xid/xid	+	−	+
bg/bg	nu/nu	+/+	−	+	−
bg/bg	+/+	xid/xid	+	−	−
+/+	nu/nu	xid/xid	−	−	+
bg/bg	nu/nu	xid/xid	−	−	−

the remainder of the immune system. As an example, Andriole et al. (1985), using some of the combinations given in Table 2-2, were able to demonstrate that NK cells function independently from lymphokine-activated killer (LAK) cells.

nu (Nude); *Dh* (Dominant Hemimelia)

Mice with the mutations nude and dominant hemimelia (*nu/nu Dh/+*) are both athymic and asplenic. They are sometimes referred to as lasat mice (Lozzio, 1976), although this is not standardized nomenclature. The immunologic deficits of *nu/nu +/+*, *+/+ Dh/+*, and *nu/nu Dh/+* mice have been described and compared (Ikeda and Gershwin, 1978; Erickson and Gershwin, 1981). Some of the deficits in the double mutant (e.g., the histologic organization of lymph nodes) are predictable. Other aspects of immune function in these animals (e.g., the response to PHA, the numbers of circulating Thy-1.2-bearing cells, and the acceptance of tumor allografts) are modified differently in the absence of both the thymus and spleen than they are in the absence of either organ alone. The *nu/nu Dh/+* mutant provides a model for investigating the functional relationship between the spleen, thymus, and bone marrow.

Yaa (Y-Linked Autoimmune Accelerator); *bg* (Beige)

An SB/Le-*sa bg/sa bg* male provided the aberrant Y-linked autoimmune accelerator gene (*Yaa*) in the creation of strain BXSB/Mp (see strain BXSB/Mp, page 96). In addition to developing the BXSB recombinant strain from offspring of C57BL/6J × SB/Le-*sa bg/sa bg Yaa* F$_2$ breeders, the segregating inbred strain SB/Le-*bg/bg Yaa* and *bg/+ Yaa* was created. It became apparent that the SB/Le-*bg/bg Yaa* males had a retarded development and progression of the clinical phase of autoimmune disease seen in the SB/Le-*bg*/+ *Yaa* males. Mice with both genotypes had similarly high levels of ANA and serum IgG, lymphadenopathy, and splenomegaly with hemolytic anemia. However, *bg/bg Yaa* males did not develop immune complex glomerulonephritis with nephrotic syndrome and survived twice as long (69 weeks) as *bg*/+ *Yaa* males (35 weeks) (Roths and Murphy, 1982). In studies by Clark et al. (1982), the findings of decreased deposition of IgM in renal glomeruli, reduced LPS responsiveness, and reduced number of ThB$^+$ spleen cells suggested that the *bg* gene could be providing a protective effect by acting directly on B cells. The role of NK cells, which are impaired in *bg/bg* mice, has not been adequately explored.

Yaa (Y-Linked Autoimmune Accelerator); *xid* (X-Linked Immune Deficiency)

The X-linked immunodeficiency (*xid*) mutation has been shown to delay the expression of accelerated autoimmunity determined by the *Yaa* gene. Studies by Golding et al. (1983) have shown that CBA/Ca × BXSB/Mp F_1-*Yaa* hybrid males, but not CBA/N × BXSB/Mp F_1-*xid Yaa* hybrid males, have progressive autoimmune lymphoproliferative disease. These latter *xid Yaa* males are devoid of splenic B colonies, are unresponsive to trinitrophenol (TNP)-Ficoll, and do not develop splenomegaly or lymphadenopathy. At 10 months of age nearly all of the CBA/Ca × BXSB/Mp F_1-*Yaa* males, but none of the age-matched CBA/N × BXSB/Mp F_1-*xid Yaa* males, are antinuclear antibody positive.

Smith et al. (1983) reported on the development of the congenic BXSB-*xid* strain. Their studies demonstrated that the *xid* mutation markedly reduced autoimmune disease in BXSB-*Yaa* males. The double mutants showed reduced lymphoid hyperplasia, hypergammaglobulinemia, and autoantibody levels; less severe renal disease; and prolonged survival. They concluded that the autoimmune disease of BXSB/Mp-*Yaa* males is dependent on B-cell subset depletion in *xid* individuals.

INBRED STRAINS OF MICE
BSVR, BSVS

Genetics

In 1930 Webster developed, from Rockefeller Institute stock, lines of mice that were resistant (BR) and susceptible (BS) to *Bacillus enteritidis* (*Salmonella* spp.) (Webster, 1933). Later tests with louping ill and St. Louis encephalitis viruses in these lines led to the development of virus-resistant and -susceptible derivatives (VR and VS) and thus to the strains BRVR,

BSVS, BRVS, and BSVR (Webster, 1937). Recent evidence indicates that the bulk of the susceptibility of the BSVS mouse to *Salmonella* is controlled by a single autosomal locus unrelated to *H-2*, *Igh-C*, or *Hbb* (Briles et al., 1977, 1981). Using F1 hybrids, O'Brien et al. (1981) have provided evidence that the genetic defect in this strain is probably due to *Ity^s*. Thus, many of the differences in the BR and BS strains are probably due to genetic differences at the *Ity* locus. Interestingly, the BSVR mouse has been found to be much more susceptible to salmonella infection than is the BSVS mouse (Benjamin and Briles, 1982). This added susceptibility is probably due to an additional non-*Ity* gene or genes. It is not known whether this gene has any relationship to the viral resistance or susceptibility traits of this group of mouse strains.

The BRVR and BRVS strains are now extinct and will not be discussed further.

Pathophysiology

BSVR and BSVS differ in their susceptibility to experimental autoimmune encephalomyelitis (Olitsky and Lee, 1953) and thyroiditis (Rose et al. 1973). A defect in thymus-dependent antibody responses in BSVS mice has been suggested (Briles et al., 1979). Streptococcal group A carbohydrate (GAC), when injected as a component of a killed streptococcal vaccine, is a T-cell-dependent antigen (Briles et al., 1982). BSVS mice make a poor immune response to GAC, apparently because of regulation by genes linked to the *H-2* and *Igh-C* loci, as well as to an additional gene or genes (Briles et al., 1977).

It was originally thought that the poor response of BSVS mice to T-cell-dependent antigens might account for the susceptibility of this strain to *Salmonella* infection (Briles et al., 1979). However, in 1981 Briles et al. demonstrated that a *Salmonella* resistance gene in A/J mice (presumably *Ity^s*) rendered these mice resistant to *Salmonella* infection without conferring the high T-cell-dependent immune responsiveness of A/J mice.

The mechanism by which the *Ity* locus affects the susceptibility to *S. typhimurium* is in dispute. It has been claimed that the locus regulates both the efficiency with which *S. typhimurium* are killed in macrophages in vitro (Lissner et al., 1983) and their rate of multiplication within the host (Hormaeche, 1980). It seems likely that the latter claim correctly represents the in vivo circumstances.

Survival of mice following infection with *S. typhimurium* is thought to be largely the result of the activation of macrophages by either LPS or T cells, so that they can more effectively kill intracellular *S. typhimurium* (Collins and Mackaness, 1968; O'Brien et al., 1980). The findings that the effects of the *Ity* locus can be observed in *nu/nu* mice (lacking T cells) (O'Brien and Metcalf, 1982) and Lps^d mice (macrophages unresponsive to LPS) (Briles et al., 1986) indicate that the effect of *Ity* on resistance to *Salmonella* is not by the modulation of either of these macrophage activation mechanisms. Furthermore, *S. typhimurium* with full virulence properties, but lacking a functional *aroA* gene (which is necessary for the synthesis of aromatic compounds that are absent in vivo), fails to show significant growth in vivo (Hoiseth and Stocker, 1981). When this organism is injected into a panel of BXD recombinant inbred mice, the number of *aroA Salmonella* recoverable from the mice after several days is totally unrelated to the *Ity* type of the mice, a finding that indicates that the *Ity* locus does not effect killing in vivo (Benjamin and Briles, 1982).

This conclusion has been confirmed by infecting mice with *S. typhimurium* carrying a plasmid that fails to replicate at 37°C. The dilution of the plasmid in the *Salmonella* taken from infected mice allows the rate of in vivo growth of the *Salmonella* to be calculated. The total recovery of the plasmid from the mice provides an unbiased index of killing. The data obtained by this procedure make it clear that the major effect of the alleles of the *Ity* locus in vivo is to modulate *Salmonella* growth (Benjamin et al., 1987).

Husbandry

Special procedures are not required for maintaining BSVR and BSVS mice.

Reproduction

BSVS mice reproduce very poorly when they are raised under conventional conditions. The major problem is high neonatal mortality. Necropsy examination of offspring reveals the absence of milk in the stomach. Such observations are consistent with murine coronavirus (mouse hepatitis virus) infection. In fact, when this strain is reared under pathogen-free conditions, the mice have larger litters and neonatal survival is much improved.

The frequency of stillborn litters increases with age in BSVR females. Therefore, it is recommended that they be bred as soon as they reach sexual maturity (D. E. Briles, University of Alabama, Birmingham, unpublished data).

BXSB/Mp Females

Genetics

BXSB/Mp is a recombinant inbred strain of mice derived from a cross between a C57BL/6J female and a SB/Le male. The inbred strain SB/Le is homozygous for the linked mutations satin (*sa*) and beige (*bg*). Both the *sa*

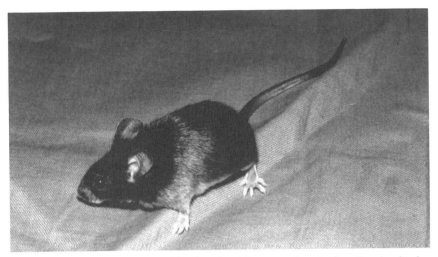

The BXSB mouse, a recombinant inbred strain with an agouti coat color. Females develop autoimmune disease during their second year of life. Males carry the Y-linked autoimmune accelerator (*Yaa*) gene and develop autoimmune disease in their first year of life. Photograph courtesy of Fred W. Quimby, New York State College of Veterinary Medicine, Cornell University, Ithaca, New York.

and *bg* genes were removed from the developing BXSB strain by negative selection. Initial aging studies demonstrated a major sex difference in life span (Murphy and Roths, 1978a). Accelerated autoimmune disease in males is due to the presence of the Y-linked gene *Yaa* described previously (see page 87). SB/Le females also develop autoimmune disease. Presumably, both the Y-linked autoimmune accelerator gene (*Yaa*) and an autosomal gene or genes responsible for induction of autoimmunity in BXSB/Mp mice were derived from strain SB/Le.

Pathophysiology

Female BXSB/Mp mice develop lymphoid hyperplasia, splenomegaly, and Coombs'-positive hemolytic anemia; however, the age at onset is two to three times greater than that of their male siblings. The serologic and immunologic abnormalities characteristic of male *BXSB-Yaa* mice by 5 months of age also occur in older (greater than 12 months of age) BXSB females. BXSB females survive to a mean age of 15 months. At necropsy the common findings are severe, chronic immune complex glomerulonephritis and vascular disease.

Husbandry

Special husbandry procedures are not required.

Reproduction

BXSB/Mp females give birth to an average of five pups per litter and in paired matings will produce only two to three litters because of the early morbidity of males.

DBA/2Ha

Genetics

DBA/2Ha was established as a subline of DBA/2 in 1933 (Bailey, 1979).

Pathophysiology

B cells of the DBA/2Ha strain fail to respond to some T-cell-replacing factors (Tominaga et al., 1980). The precise nature of the defect and its genetics is uncertain. Takatsu and Hamaoka (1982) concluded that an X-linked recessive allele was responsible for a failure to respond to activation factors produced by activated T cells and the T-cell hybridoma B151K12. It is likely that the factor in question is B-cell growth factor type 2 (IL-6) (Takatsu et al., 1985). Antibody against the putative receptor has been prepared (Tominaga et al., 1980). Sidman et al. (1986a), using a different assay system, concluded that the defect caused a lack of response to γ-interferon and to two other B-cell maturation factors (Bmf). They also reached the conclusion that the genes mapped to autosomal loci (*Bmfr-1* on chromosome 4 and *Bmfr-2* on chromosome 9). It remains unresolved whether the differences in these experiments are technical or are due to a segregation of genes within the DBA/2 lineages.

Affected mice do not appear to be susceptible to overt infection or to have immune deficiencies.

Husbandry

Special husbandry procedures are not required.

Reproduction

These animals reproduce normally.

MRL/Mp

Genetics

The MRL/Mp strain was developed as a by-product of a series of crosses involving AKR/J, C57BL/6J, C3H/Di, and LG/J mice for the purpose of creating a compatible genetic background for maintaining the mutation achondroplasia (*cn*). It is estimated that 75 percent of the MRL/Mp genome is derived from strain LG/J. The mutant gene lymphoproliferation (*lpr*) arose during the development of this inbred strain (see page 59) (Murphy and Roths, 1977).

Nearly all MRL/Mp mice have ANAs by 5 months of age. By contrast, no ANAs have been detected in LG/J mice as old as 24 months. To examine the genetics of this phenomenon, F1 and F2 hybrid matings and F1 backcross matings to MRL/Mp were performed; and the offspring were assayed for ANAs at 4 months of age. The data from the F1 and F2 hybrid matings suggest that the MRL strain contributed a single autosomal recessive gene. However, the F1 backcross to MRL produced a greater than expected frequency of ANA-positive offspring. More work must be done to determine whether a single gene causes the development of ANA in MRL/Mp mice (Roths, 1987).

Pathophysiology

MRL/Mp mice develop clinically defined autoimmune disease without lymphoid hyperplasia. By 4–5 months of age the sera of 94 percent of MRL/Mp mice have ANAs. These mice also spontaneously produce anti-DNA and anti-ribonucleoprotein (RNP) autoantibodies, and anti-Sm is found in 35 percent of males and 45 percent of females (Billings et al., 1982). Retroviral gp70 immune complexes are found in MRL/Mp females over 1 year of age (Izui et al., 1979). Nearly all MRL/Mp mice have severe chronic glomerulonephritis resembling the lupuslike renal disease of NZB × NZW F_1 hybrid females. MRL/Mp mice have extremely elevated levels of urinary protein after 14 months of age (Kelley and Roths, 1985). Degenerative arterial disease and polyarteritis are also common. MRL/Mp mice show widespread inflammatory infiltrates involving cerebral vessels and meninges but not the choroid plexus (Alexander et al., 1983). At autopsy, one-half have malignant tumors, one-third of which are reticulum cell neoplasms. The mean life span of MRL females and males is 74 and 92 weeks, respectively (Murphy, 1981). Early exposure to physiologic levels of estrogen modifies humoral components of the disease in both MRL +/+ and MRL−*lpr/lpr* mice (Brick et al., 1988).

Husbandry

MRL/Mp mice are large and appear to be healthy during their breeding period. Although special husbandry procedures are not required, these animals are maintained at the Jackson Laboratory, Bar Harbor, Maine, on a diet of pelleted 96W chow (21.9 percent protein, 7.2 percent fat).

Reproduction

MRL/Mp mice are excellent breeders. In one study of reproductive performance, 37 breeding pairs produced an average of 4.5 litters per pair, with an average of 6.2 pups per litter and 96 percent of the pups surviving to weaning (Murphy and Roths, 1978c).

NOD (Non-Obese Diabetic)

Genetics

NOD (non-obese diabetic) is an inbred strain of mice derived from ICR/Jcl mice by selection for spontaneous development of insulin-dependent diabetes (Idd) (Makino et al., 1980). A polygenic basis for diabetes susceptibility in the NOD strain has been established by outcrossing to the related, diabetes-resistant inbred strain NON (non-obese normal), which was separated by Makino from the NOD line at the sixth generation of inbreeding. For diabetes to occur following outcrossing to NON and backcrossing to NOD, a minimum of three recessive diabetogenic genes must be inherited in the homozygous state from NOD (Prochazka et al., 1987). The first recessive diabetogenic locus, observed initially by Hattori et al. (1986) in mice outcrossed to C3H and backcrossed to NOD, is tightly linked to the MHC on chromosome 17 and has been provisionally designated *Idd-1* (Leiter et al., 1986). It has not yet been definitively shown, however, that this locus is within the MHC. A unique *I-Aβ* locus (Acha-Orbea and McDevitt, 1987) and a failure to express an *I-E* gene product (Nishimoto et al., 1987) have also been proposed. The second diabetogenic locus, *Idd-2*, is on chromosome 9 (Prochazka et al., 1987), about 15 cM centromeric to the *Thy-1/Alp-1* marker loci (M. Prochazka, D. V. Serreze, S. M. Worthen, and E. H. Leiter, The Jackson Laboratory, Bar Harbor, Maine, unpublished data). The third locus has not yet been mapped but can be shown to segregate in a second backcross (BC2) to NOD using diabetes-free BC1 individuals typed for homozygosity for the NOD alleles at marker loci linked to *Idd-1* and *Idd-2*. In addition to these recessive genes, a gene or genes underlying T-lymphocyte proliferation is inherited from NOD in a dominant fashion. Segregation analysis using a NOD.NON-*H-2* congenic stock has shown that the increased percentage of T lymphocytes in NOD is not MHC linked (M. Prochazka, D. V. Serreze, S. M. Worthen, and E. H. Leiter, The Jackson Laboratory, Bar Harbor, Maine, unpublished data). In an outcross of NOD to C57BL/10J followed by a backcross to NOD, Wicker et al. (1987) found that the development of diabetes and insulitis is under partially overlapping but distinct genetic control, with the initiation of insulitis determined by a single gene unlinked to the MHC. However, the essential role of *Idd-1* in diabetes pathogenesis is indicated by the finding that no diabetes and practically no insulitis could be observed in NOD.NON-*H-2^b* congenic homozygous mice produced at the fifth backcross (M. Prochazka, D. V. Serreze, S. M. Worthen, and E. H. Leiter, The Jackson Laboratory, Bar Harbor, Maine, unpublished data).

A further indication of the overlapping control of insulitis by MHC-linked genes has been provided by a transgenic mouse study. NOD mice do not express I-E surface antigen. C57BL/6-*Eα^d* transgenic mice, which do express I-E surface antigen, were mated to NOD mice, and the F1 progeny that expressed I-E surface antigen were backcrossed with NOD to produce offspring differing in I-A and I-E phenotypes (and segregating for other *Idd* loci). It was found that the expression of I-E surface antigen in backcrossed mice homozygous for the NOD *I-A* marker gene prevented the appearance of insulitis at 9 weeks of age (Nishimoto et al., 1987). To assess the effect of this transgene on diabetes development, it must be injected directly into NOD zygotes since outcrossing NOD with C57BL/6J strains produces very low diabetes incidence at BC1 (L. Herberg, Diabetes Institute, Dusseldorf, Federal Republic of Germany, personal communication to E. H. Leiter, 1988). The introduction of single mutant genes associated with glucose intolerance syndromes in mice (*db*, *ob*, *A^y*) on the NOD background produced transitional hyperplasia of islets followed by insulitis, B-cell atrophy, and severe diabetes characteristic of NOD (Nishimura and Miyamoto, 1987). The genetic, pathologic, and therapeutic implications of diabetes in NOD mice have been reviewed (Tochino, 1987).

Pathophysiology

Clinical features of the diabetes syndrome in NOD mice are quite similar to human type I diabetes, including abrupt onset between 90 and 120 days of age (equivalent to early adolescence in humans), ketonuria, glycosuria, hyperglycemia, hypercholesterolemia, polydipsia, polyuria, and polyphagia. An autoimmune etiopathogenesis is indicated by the infiltration of T lymphocytes into degenerating pancreatic islets (insulitis), which is detectable shortly after weaning (Miyazaki et al., 1985); production of spontaneous autoantibodies against islet antigens (Kanazawa et al., 1984); and lymphoid tissues unusually enriched for T lymphocytes. Heavy lymphoid cell aggregations are also found in the salivary (submandibular) gland in both sexes (Makino et al., 1985).

Females develop diabetes at an earlier age and with a higher frequency than males. Ovariectomy reduces and orchiectomy increases the incidence of diabetes (Makino et al., 1981). The incidence of diabetes in females is approximately 80 percent in most colonies; the incidence in males varies between 10 and 70 percent. The wide variation in incidence among males is thought to be due to environmental factors, including environmental pathogens, that can modulate disease expression. This has been confirmed by using NOD males from a colony in which the incidence of diabetes in males was below 10 percent. Producing these animals under germfree conditions caused the incidence to rise to 70 percent (Suzuki et al., 1987). While germfree conditions accelerate diabetes, a variety of systemic immunostimulatory treatments, including viral infection (Oldstone, 1988) and injection of bacterial cell wall preparations (Toyota et al., 1986), prevent diabetes. Apparently, NOD mice are capable of generating immunoregulatory protective mechanisms against autoreactive cells when their immune systems are stimulated by antigen challenges in the environment.

NOD-nu/nu mice do not develop diabetes unless they are reconstituted with splenocytes from euthymic littermates and concomitantly treated with IL-2, indicating a requirement for T lymphocytes to mediate pathogenesis (Makino et al., 1986). Moreover, transferring splenocytes from older NOD mice into irradiated 7-week-old recipients accelerates the development of diabetes in the younger animals (Wicker et al., 1986). Using this syngeneic transfer paradigm, others have shown that both L3T4$^+$ and Ly-2$^+$ T-cell subsets are necessary to mediate β-cell destruction (Bendelac et al., 1987; Miller et al., 1988). Accordingly, treatment of prediabetic NOD mice with a wide variety of immunosuppressive reagents that deplete T-cell subsets prevents the development of overt diabetes. For example, Shizuru et al. (1988) have shown that sustained treatment of NOD mice with monoclonal antibody directed against the L3T4 determinant of murine Th cells halts the progression of diabetes and, in some cases, leads to long-term reversal of the disease after therapy is discontinued.

NOD bone marrow cells can transfer diabetes to irradiated diabetes-resistant F1 hybrids, showing that the polygenes controlling diabetes susceptibility must be expressed in marrow-derived effector cells (Serreze et al., 1988c; Wicker et al., 1988). NOD mice exhibit a spectrum of immunoregulatory defects. Both monocyte and NK cell functions are impaired (Kataoka et al., 1983; Serreze and Leiter, 1988), and T cells generated in a syngeneic MLR fail to induce suppression of an MLR (Serreze and Leiter, 1988). The defective ability of antigen-presenting cells to activate T-suppressor inducer cells has been associated with cytokine deficiencies, including both decreased endoxin-stimulated IL-1 production from NOD macrophages and decreased endogenous production of IL-2 from T lymphocytes in syngeneic MLRs (Serreze and Leiter, 1988). Functional T-suppressor inducer blast cells can be elicited in vitro by exposure to ConA, IL-1, or IL-2 (Serreze and Leiter, 1988). NOD mice treated in vivo with low doses of IL-2 not only exhibit a normal syngeneic MLR but also show markedly reduced diabetes incidence (Serreze et al., 1988a). NOD/Lt mice that are free of the diabetic syndrome by 1 year of age are prone to develop follicle center cell lymphomas.

Husbandry

Insulin treatment is required to maintain diabetic mice; without insulin they survive only 1–2 months after diagnosis. Diabetes is diagnosed by determination of an elevated (nonfasting) blood or plasma glucose level. This determination can be made by measuring blood glucose directly or by screening for glycosuria using Tes-Tape® (Eli Lilly & Co.). Glycosuria, as read by the Tes-Tape®, usually denotes a plasma glucose of 300 mg/dl. Large numbers of mice can be easily screened by this method.

Diet appears to be extremely important in the expression of diabetes. Feeding NOD mice Pregestimil® (a nonallergenic infant formula containing a casein hydrolysate profile of amino acids) does not support the same level of diabetes as does a semidefined diet containing a broad spectrum of ingredients (Elliott et al., 1988). At the Jackson Laboratory, Bar Harbor, Maine, a high incidence of diabetes in both males (≥ 50 percent) and females (≥ 90 percent) is obtained by maintaining the colony on a diet of pelleted Old Guilford 96W, a complex formulation containing wheat germ, ground wheat, soybean meal, and brewer's yeast as well as milk proteins. Feeding the chemically defined diet AIN-76 to mice from the same colony delays the development of diabetes by 3–4 months. The addition of 25 percent (w/w) of Old Guilford 96 to AIN-76 will reestablish the expected onset of diabetes (D. L. Coleman, The Jackson Laboratory, Bar Harbor, Maine, personal communication to E. H. Leiter, 1988). The presence of immunosuppressive compounds in the complex grain-containing diets is suspected.

Reproduction

NOD mice are maintained by brother × sister mating. They have an excitable disposition but breed well, even though they have been inbred for more than 40 generations. Siblings bred before the development of overt diabetes can usually produce two large (9–14 pups) litters in which nearly all the pups survive to weaning. Breeders can be protected from developing diabetes by a single injection of complete Freund's adjuvent (B. Singh, Department of Immunology, University of Alberta, Edmonton, Alberta, Canada, unpublished data).

NON (Non-Obese Normal)

Genetics

Non-obese normal (NON) mice were separated from the NOD line at the sixth generation of inbreeding (Makino et al., 1985). Paradoxically, NON progenitors were initially selected for a high fasting blood glucose level with the goal of producing a spontaneously diabetic model, while the progenitors of the NOD strain were initially selected for a normal fasting blood glucose level to provide a nondiabetic control. However, as described in the preceding section, spontaneous autoimmune diabetes developed in the NOD line, and NON mice, despite the strain name, are neither normal nor nonobese.

Pathophysiology

NON males exhibit impaired glucose tolerance after weaning (Tochino et al., 1983) and become obese (50 g) by 20 weeks of age (E. H. Leiter, The Jackson Laboratory, Bar Harbor, Maine, unpublished data). Although islet morphology and β-cell granulation remain normal, the strain is characterized by the development of severe kidney lesions (Tochino et al., 1983). NON/Lt mice also exhibit anomalies in their immune systems. T-lymphocyte number and function are normal in young NON/Lt mice, but T lymphocytopenia and lymphopenia, coupled with declining T-cell mitogen responses, develop in these mice as they age (Leiter et al., 1986). NON/Lt mice are similar to NOD/Lt mice in that they exhibit a defective syngeneic MLR, but unlike NOD/Lt mice, which lack functional NK cells, NON/Lt mice exhibit normal NK cell activity levels (D. V. Serreze and E. H. Leiter, The Jackson Laboratory, Bar Harbor, Maine, unpublished data). The unique features of the MHC of NON mice and another related strain, CTS, have been described recently (Ikegami et al., 1988). Although NON/Lt mice are distinguished from NOD/Lt mice by the absence of insulitis, leukocytic infiltrates of the submandibular gland are a histopathologic characteristic shared by both strains. Serum autoantibodies against insulin and p73, an endogenous retroviral protein, have been detected by enzyme-linked immunosorbent assay (ELISA) in NON/Lt mice, although titers are considerably lower than those in NOD/Lt mice (Serreze et al., 1988b).

Husbandry

Special husbandry procedures are not required.

Reproduction

These animals reproduce normally; however, litter sizes are smaller (5–8 pups) than those of NOD mice (9–14 pups).

NZB (New Zealand Black)

Genetics

The NZB strain was produced by inbreeding from an outbred mouse colony and selecting for black coat color (Bielschowsky et al., 1956). The NZB

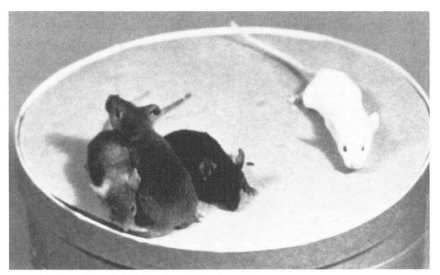

NZB, NZW, and BWF$_1$ hybrid mice. NZB mice (right center, black coat) develop hemolytic anemia; NZW (right, white coat) and BWF$_1$ hybrid mice (left and left center, agouti coats) develop a systemic lupus erythematosus-like disease. Photograph courtesy of Fred W. Quimby, New York State College of Veterinary Medicine, Cornell University, Ithaca, New York.

strain and the NZB × NZW F$_1$ hybrid strain have been used as prototype strains for the study of spontaneous autoimmune disease. Studies with these strains were designed to elucidate the genetic basis of autoimmunity and have demonstrated complex polygenic inheritance (Theofilopoulos and Dixon, 1985). NZB mice develop autoimmune hemolytic anemia at an early age (Bielschowsky et al., 1959) and, in addition, have elevated levels of immunoglobulin, anti-DNA antibodies, anti-thymocyte antibodies, and circulating immune complexes (Quimby and Schwartz, 1982; Theofilopoulos and Dixon, 1985). Genes that contribute to these abnormalities in mice have been postulated: three genes (*Aia-1*, *Aia-2*, *Aem-1*) controlling anti-erythrocyte antibody production, two genes (*Nta-1*, *Ntm-1*) controlling anti-thymocyte antibody production, four genes (*Ass-1*, *Ass-2*, *Ass-3*, *Ass-4*) controlling production of anti-ssDNA, four genes (*Ads-1*, *Ads-2*, *Ads-3*, *Ads-4*) controlling production of anti-dsDNA, one gene (*Imh-1*) controlling elevated IgM, and two genes (*Agp-1*, *Agp-3*) controlling levels of anti-gp70 circulating immune complexes. The proposed genotype of the NZB strain is *Aia-1*$^+$, *Aia-2*$^+$, *Aem-1*$^-$, *Nta-1*$^+$, *Ntm-1*$^+$, *Ass-1*$^+$, *Ass-2*$^+$, *Ass-3*$^-$, *Ass-4*$^-$, *Ads-1*$^+$, *Ads-2*$^+$, *Ads-3*$^-$, *Ads-4*$^-$, *Imn-1*$^+$, *Agp-1*$^+$, *Agp-3*$^-$. Four of these genes, *Ads-1*, *Ass-1*, *Agp-1*, and *Nta-1*, are linked to the H-2^d haplotype (MHC class I genes) on chromosome 17 (Shirai et al., 1984).

Pathophysiology

The most consistent abnormality seen in NZB mice is Coombs'-positive hemolytic anemia. Anti-erythrocyte antibodies appear as early as 3 months of age and reach an incidence of 100 percent by 12–15 months of age. Initially, these antibodies are of the immunoglobulin IgG class, in contrast to those observed later in life, which are of the IgM class.

Hemolytic anemia usually develops 5 months following the appearance of autoantibodies and is not gender specific. Anemia is associated with reticulocytosis and reduced erythrocyte survival time. Splenomegaly is present as a result of erythrocyte sequestration, increased hematopoiesis, and lymphoid hyperplasia. Both males and females have an average life span of 16.6 months (Eastcott et al., 1983). Andrews et al. (1978) reported a mortality of 90 percent by 23 months.

Lymphoproliferative lesions resulting in hyperplasia of spleen, lymph nodes, bone marrow, thymus, lung, kidney, liver, and salivary glands are consistent features of the disease process in NZB mice. Two phases of lymphoproliferation are seen. Between 3 and 11 months of age, the white pulp of the spleen and both cortical and medullary regions of the lymph nodes are characterized by enlarged lymphoid follicles containing multiple germinal centers. Later in life a second phase of lymphoproliferation occurs; this is characterized by extreme plasma cell hyperplasia in lymphoid tissue throughout the body. An increased incidence of lymphoma has been reported (East, 1970). The thymus is characterized by hyperplasia with follicular aggregates of lymphocytes and mast cells in the medulla. There is premature thymic involution in which degeneration and vacuolization of epithelial cells are consistent features (Andrews et al., 1978). An impressive early decline in thymulin levels has been reported (Bach et al., 1973).

NZB mice begin to produce anti-DNA antibodies by 2 months of age. High anti-ssDNA titers are observed in 54 percent of NZB mice by 9 months of age; however, only 10 percent have a significant titer of anti-dsDNA antibodies at this time (Andrews et al., 1978). Natural thymocytotoxic antibodies (NTAs) are made by all NZB mice by 7 months of age (Maruyama et al., 1980). This autoantibody is cytotoxic for thymocytes and T cells and reacts with brain tissue. NTAs are known to react with the Thy-1 complex as well as with a T-cell differentiation antigen expressed during the early stage of T-cell maturation (Surh et al., 1987). T cells with suppressor cell function appear to have the highest density of NTA-reactive antigen on their surface (Shirai et al., 1978), and NTA has been postulated in one laboratory as the cause of the T-cell abnormalities (Shirai et al., 1972). However, Taurog et al. (1981) demonstrated that the T-cell defects seen in NZB mice are also seen in NZB.CB/N-*xid* mice, which do not produce NTA, suggesting a primary T-cell defect in the NZB strain. A primary T-cell defect was also proposed by Laskin et al. (1986) based on cell mixing experiments and by Miller and Calkins (1988), who demonstrated an active role for T cells in promoting an in vitro autoantibody response against erythrocytes.

A primary B-cell defect has also been clearly demonstrated in NZB mice. This hyperactivity of B cells is manifested by high levels of IgG immunoglobulins by 3 months of age (Andrews et al., 1978), increased number and augmented secretion of IgM by B cells (Manny et al., 1979), and the production of numerous autoantibodies (Quimby and Schwartz, 1982). Defective clonal inactivation of autoreactive B cells has been proposed to account for the increased autoantibodies seen in NZB mice (Cowdery et al., 1987). However, Cantor et al. (1978) provided evidence that there was impaired feedback regulation of antibody synthesis because of an abnormally functioning Ly-123$^+$ T-cell subset. At least one B-cell defect is known to reside in a Lyb-5$^+$ subpopulation of B lymphocytes (Ly-1$^+$ B cell) characterized by the normal allele of the *xid* mutant gene (A. D. Steinberg et al., 1982). This Ly-1$^+$ B cell is increased in young NZB mice, is responsible for much of the autoantibody produced, and has unusual oncogene and receptor gene expression (A. D. Steinberg et al., 1987; Wolfsy and Chiang, 1987). NZB splenic B cells, which are rich in Ly-1$^+$ B cells, are capable of transferring autoantibody production into NZB-*xid* mutant mice (Ishigatsubo et al., 1987), which again suggests that a primary defect resides in a subpopulation of NZB B cells. However, others have demonstrated that autoantibodies and autoimmune disease can occur independent of B-cell hyperactivity (putative Ly-1$^+$ B cells) in NZB mice backcrossed to other strains (Datta et al., 1982; Eastcott et al., 1983). Furthermore, abnormal B cells are observed in athymic (nude) mice (Ohsugi and Gershwin, 1979). The B lymphocytes of NZB mice appear to be more resistant to induction of immunologic tolerance relative to those of other mice. It has been shown that NZB splenic B cells require a higher epitope density of the trinitrophenyl hapten on a protein carrier to induce unresponsiveness. The mechanism for this increased B-cell resistance to induction of unresponsiveness was independent of T cells, macrophages, or a Fcγ receptor defect (Goldings, 1988).

The expression of infectious xenotropic retrovirus in NZB mice is controlled by two genes, *Nzv-1* and *Nzv-2* (Datta and Schwartz, 1976, 1977). In addition to high levels of gp70 envelope antigen in the blood, NZB mice also have high levels of anti-gp70 immune complexes in their serum (Andrews et al, 1978). However, the incidence of glomerulonephritis in NZB mice is low (Knight and Adams, 1978). Datta et al. (1978b) have shown that the genes controlling virus expression are independent of the development of autoantibodies or glomerulonephritis.

Transplantation of bone marrow from autoimmune-resistant mice to NZB mice corrects all the abnormalities in immunity and prevents the development of autoimmune disease. Likewise, transplantation of bone marrow from NZB mice to young autoimmune-resistant mice results in both immunologic abnormalities and autoimmune disease (Akizuki et al., 1978; Jyonouchi et al., 1981).

Husbandry

Spontaneous infection of NZB mice with certain murine viruses has been shown to modify the course of the autoimmune disease (Tonietti et al., 1970); therefore, this strain should be maintained in a pathogen-free environment.

Reproduction

This strain breeds normally.

NZB × NZW F_1 Hybrids

The NZW (New Zealand White) strain was produced by W. H. Hall in 1952 by inbreeding from an outbred mouse colony and selecting for white coat color (Talal, 1983). Although mice of this strain have relatively normal life spans, they do develop anti-DNA antibodies, high serum levels of retroviral gp70 antigen, and nephritis later in life (Kelley and Winkelstein, 1980).

NZB × NZW F_1 hybrids (hereafter called BWF_1) develop a systemic lupus erythematosus (SLE)-like syndrome. It has been proposed that the NZB parent contributes the dominant *Ads-1* and *Ads-2* genes controlling anti-dsDNA production, and the NZW parent contributes the dominant *Ads-3* and *Ads-4* genes, which are modifier genes. Similarly, the NZW parent contributes two dominant genes, *Ass-3* and *Ass-4*, which enhance the effect of *Ass-1* and *Ass-2* in the production of anti-ssDNA antibodies. The *Agp-3* gene of NZW intensifies the effect of the NZB *Agp-1* gene controlling anti-gp70 circulating immune complexes. However, the NZW *Aem-1* gene suppresses the activity of the NZB *Aia-1* gene responsible for anti-erythrocyte antibody production. Finally, a dominant trait coding for lupus nephritis, *Lpn-1*, is modified by two additional genes, *Lpn-2* and *Lpn-3*, both of which are donated by the NZW partner. The genes *Ass-3*, *Ads-3*, *Agp-3*, and *Lpn-2* are linked to the $H-2^z$ haplotype of NZW (Shirai et al., 1984, 1987; Bearer et al., 1986). The net result is that BWF_1 hybrids have an intensified production of anti-dsDNA antibodies, anti-ssDNA antibodies, and circulating immune complexes; an increased susceptibility to lupus nephritis; and a decreased production of anti-erythrocyte antibody when compared with NZB. Although evidence exists for each of the proposed loci listed above, there is still disagreement concerning the precise assignment of genetic loci to the autoimmune phenotype (Kotzin and Palmer, 1987). Recently Kotzin et al. (1985) demonstrated a large deletion (8.8-kilobase segment) in the DNA containing $C\beta_1$, $D\beta_2$ and $J\beta_2$ cluster encoding the T-cell β chain; this has been confirmed recently (Theofilopoulos, 1986). The functional significance of this deletion in the BWF_1 hybrid is unknown.

Pathophysiology

BWF_1 mice develop a disease characterized by high levels of antibodies directed toward nucleic acid antigens, progressive immune complex glomerulonephritis, and a marked enhancement of the disease in females. As early as 2 months of age, ANA can be detected in some BWF_1 mice. By 12 months of age all BWF_1 hybrids have detectable levels of ANA (Andrews et al., 1978; Quimby and Schwartz, 1982). The anti-dsDNA antibodies have nephritogenic properties and appear to be principally responsible for the immune complexes deposited in the glomerulus (Lambert and Dixon, 1968). Unique subsets of T and B lymphocytes are found in BWF_1 mice that are responsible for the production of pathogenic (cationic) IgG anti-DNA (Datta et al., 1987).

Hybridomas secreting anti-dsDNA antibodies derived from BWF_1 mice injected with phosphocholine were isolated, and two DNA sequences associated with the rearranged immunoglobulin heavy- and light-chain genes were compared with the VH germline sequences of normal BALB/c and C57BL/10 mice as well as the germline BWF_1 gene. All 11 hybridomas binding dsDNA used the same V_H11 germline gene and 9 used the same heavy-chain VDJ and light-chain VJ combinations. Five of these hybridomas were used for comparative purposes, and the nucleotide sequences encoding the heavy-chain variable regions differed from those of the germline by 6 to 16 base changes, indicating extensive somatic diversification. The authors speculated that the large number of base substitutions and the IgG2a subclass strongly suggest that T cells affected both the proliferation and differentiation of the B cells that produce these autoantibodies in vivo (Behar and Scharff, 1988).

Independently, Eilat et al. (1988) compared the nucleotide sequences of rearranged heavy- and light-chain genes encoding spontaneous natural anti-DNA autoantibodies from different BWF_1 mice. They found that the H chains of two anti-DNA antibodies had V_H segments belonging to two different V_H gene families, but both had similar D segments and J sequences. Furthermore, one of these IgG anti-DNA antibodies had a heavy-chain V region encoded by the V_H11 germline nucleotide sequence. In fact, the V_H nucleotide sequence of the autoantibody differed from the germline V_H11 sequence by four nucleotides, three of which occurred in the complementarity-determining regions (CDRs). These authors suggest that this relationship between a germline gene and an expressed gene, with a concentration of nucleotide changes in the CDRs, is characteristic of an antigen-selected somatic mutation and that the actual autoantigen drives this process.

Analysis of the rearranged genes encoding spontaneous anti-RNA antibodies derived from BWF_1 mice demonstrated that there are closely related V_H gene sequences between the two antibodies but no similarity to the V_H segment of the anti-DNA antibodies (Eilat et al., 1988).

Renal changes include the "lumpy-bumpy" deposits of IgG, antigen, and complement along the glomerular capillary basement membrane, as detected by immunofluorescence microscopy. This process begins between 3 and 6 months of age, causes significant proteinuria by 6–7 months of age, and is responsible for chronic renal insufficiency and death in females by 12–14 months of age and in males by 19 months of age (Andrews et al., 1978; Knight and Adams, 1978). In addition, anti-gp70 complexes can be demonstrated in damaged glomeruli (Dixon et al., 1969).

The deposition of antigen–antibody complexes in medium and small arteries of BWF_1 mice, especially in the heart, appears to lead to thrombotic and obliterative vascular changes and myocardial infarction (Accinni and Dixon, 1979).

BWF_1 hybrids have a generalized lymphoid hyperplasia involving the spleen and all lymph nodes and an increased incidence of lymphoma (East et al., 1967). Newborn hybrids have thymic medullary epithelial cell hyperplasia, increased Hassall's corpuscles, and abundant lymphoid tissue; however, severe thymic cortical atrophy occurs early in life (Andrews et al., 1978).

In young (1- to 4-month-old) BWF_1 mice, the numbers and function of lymphocytes appear to be normal. In contrast to mice of other strains, the proliferative response of lymphocytes from young BWF_1 hybrids to mitogens and allogeneic cells in vitro and their ability to induce tumor rejection appears to be enhanced (Evans et al., 1968; Gazdar et al., 1971). Likewise, young BWF_1 mice make an augmented antibody response to nearly all antigens and are refractory to tolerance induction (Goldings et al., 1980).

Older BWF_1 hybrids show marked deficiencies in cell-mediated immunity (CMI). This abnormality is characterized by decreased lymphocyte proliferation to mitogenic stimuli, impaired graft-versus-host disease, impaired skin graft rejection, and a decreased cytotoxic response of spleen cells after alloimmunization (Howie and Simpson, 1976). These abnormalities might be due to the production of NTA (Shirai and Mellors, 1971) or abnormal differentiation of T-cell subpopulations due to lowered levels of thymic humoral factor (Bach et al., 1973). Old BWF_1 mice have an expanded population of $L3T4^-/Ly-2^-$ $Thy-1^+$ cells that can induce pathogenic autoantibody (Datta et al., 1987).

The augmented antibody response to antigens, increased levels of IgM and IgG in blood, and production of many autoantibodies suggest hyperactivity of the B-cell system. This hyperactivity of B cells occurs late in fetal life and is present at birth (Jyonouchi and Kincade, 1984). Studies by Manny et al. (1979) suggest that these B cells have a reduced number of μ-chain surface receptors when compared with γ-chain receptors and that the individual B cells are capable of synthesizing greater amounts of IgM antibody. Convincing support for a primary B-cell defect responsible for autoimmunity in BWF_1 mice was provided by Gershwin et al. (1980), who observed all of the features of autoimmunity and autoimmune disease in nude BWF_1 hybrid mice.

The defect of BWF_1 hybrid mice, like that of the NZB parent, probably resides in the bone marrow stem cell, because the immune disorders are preventable by early bone marrow transplantation (Ikehara et al., 1987). Bone marrow transplantation decreases the concentration of autoantibodies and circulating immune complexes, abrogates the development of morphologic abnormalities in the kidney and thymus, and restores cell-mediated immune function in old BWF_1 mice (Ikehara et al., 1987). As a result, bone marrow transplants from autoimmune-resistant donors prevent the expression of autoimmune disease in BWF_1 hybrids, which suggests that the etiopathogenesis of autoimmune disease lies in a primary stem cell defect.

A defective production of the hormone IL-2 has also been described in BWF_1 hybrid mice (Altman et al., 1981), and elevated levels of type II interferon (γ-interferon) have been implicated in the pathogenesis of the disease in BWF_1 mice (Engleman et al., 1981).

The greater incidence and earlier onset of autoimmune disease in female BWF_1 hybrids suggest a role for sex hormones in the pathogenesis of murine lupus erythematosus. Indeed, when male BWF_1 mice are castrated, they develop the female pattern of the disease. Likewise, androgen administration greatly prolongs the life of female BWF_1 hybrids (Raveche et al., 1979).

A variety of therapeutic measures has proven to be successful in diminishing the severity of BWF_1 lupus. Among these are immunosuppressive drugs (Jones and Harris, 1985; Dueymes et al., 1986), protein and caloric restriction (Fernandes et al., 1976; Johnson et al., 1986), a ConA-induced soluble suppressor (Krakauer et al., 1976), antibodies against γ-interferon (Jacob et al., 1987), prostaglandin E_1 (Zurier, 1982), ribavirin (Klassen et al., 1977), and an eicosapentenoic acid (fish oil)-supplemented diet (Prickett et al., 1981).

Husbandry

While certain infectious agents have been shown to induce autoantibodies and immune complex disease in normal mice (Schulman et al., 1964; Barnes and Tuffrey, 1967; Dixon et al., 1969), other infectious agents have been shown to inhibit autoantibodies and ameliorate disease in NZB and BWF_1 hybrid mice (Oldstone and Dixon, 1972). Engleman et al. (1981) proposed that virus-induced type 1 interferon is responsible for the accelerated autoimmune disease seen in some BWF_1 hybrids. Therefore, it appears prudent to maintain this hybrid strain in a pathogen-free environment.

Rabin (1985) reported a significant difference in the survival of BWF_1 hybrids associated with the cage type. Mice held in wire mesh cages lived considerably longer that F1 hybrids housed in solid-bottom cages.

Reproduction

Both parental strains breed normally; therefore, production of F1 hybrids from these parental strains requires no special practices.

NZB × SWR F_1 or SWR × NZB F_1 Hybrids

Genetics

Hybrid mice derived from the mating of mice of the autoimmune-prone NZB and normal SWR strains (hereafter called SNF_1) have been shown to develop severe early-onset glomerulonephritis and circulating anti-DNA antibodies (Datta et al., 1978a,b). Female SNF_1 mice produce nephritogenic autoantibodies encoded by the SWR parent. Immunoregulatory defects include two subpopulations of helper T cells capable of inducing nephritogenic anti-DNA antibodies from SNF_1, but not SWR, B cells. Genes found on chromosomes 6 and 17 of the normal SWR parent interact with NZB-derived genes, leading to the development of accelerated and severe nephritis in the SNF_1 mouse (Ghatak et al., 1987).

Pathophysiology

The NZB parent of the SNF_1 hybrid spontaneously develops autoimmune hemolytic anemia, expresses high levels of retroviruses and retroviral gp70 antigen, and infrequently develops glomerulonephritis (see page 104). The SWR parent, by contrast, does not express retroviruses, has very low serum gp70 levels, and does not develop autoimmune disease or circulating autoantibodies (Datta et al., 1978a,b; Eastcott et al., 1983). Female SNF_1 mice develop high serum levels of anti-DNA antibodies and lethal glomerulonephritis by 1 year of age (Datta et al., 1978a; Eastcott et al., 1983). Since severe glomerulonephritis can occur in SNF_1 females in the absence of glomerular deposits of gp70, much attention has focused on the nature of circulating and kidney-bound anti-DNA antibodies in this strain.

Analysis of monoclonal anti-DNA antibodies derived from SNF_1 and NZB mice demonstrated that most SNF_1-derived anti-DNA antibodies were of the IgG class and were cationic, compared with NZB-derived anti-DNA antibodies, which were mostly of the IgM class and were anionic (Gavalchin et al., 1985). Among SNF_1-derived anti-DNA antibodies are a set of highly cationic, IgG2b anti-DNA antibodies containing the SWR allotype and heavy-chain V regions responsible for the charge characteristics (Gavalchin et al., 1985). Since cationic IgG anti-DNA antibodies are selectively deposited in the glomerular lesions of mice with lupus nephritis (Dang and Harbeck, 1984), it was proposed that these anti-DNA antibodies were responsible for glomerulonephritis in SNF_1 mice.

Gavalchin et al. (1987) developed a library of 15 anti-idiotypic antibodies prepared by immunizing rabbits with 15 monoclonal anti-DNA antibodies derived from NZB or SNF_1 mice. Using these anti-idiotypic antibodies, they identified 10 cross-reactive idiotype (CRI) families among the 65 monoclonal anti-DNA antibodies. Five CRI families were restructured to cationic anti-DNA antibodies exclusively of SNF_1 origin and encoded by genes derived from the SWR parent (determined by allotype). These cationic anti-DNA CRI families were grouped into an interrelated cluster, Id564, which was prominently represented in the kidneys of SNF_1 mice with early nephritis (Gavalchin and Datta, 1987). Additional anti-DNA antibodies deposited in the kidneys of SNF_1 mice were identified as a second interrelated cluster, Id512, in the CRI family. These autoantibodies were not restricted to a particular charge or allotype. Both Id564 and Id512 antibodies could be found in the serum of old SNF_1 mice but not in the serum of either parental strain, suggesting that these nephritogenic idiotypes were dormant in the NZB and SWR parents and became deregulated and expanded in SNF_1 hybrids (Gavalchin and Datta, 1987).

Studies directed at elucidating an immunoregulatory defect present in SNF_1 mice showed that Th cells were essential for inducing B cells to produce highly cationic, IgG anti-DNA antibodies in vitro (Datta et al., 1987). These Th cells appear in the spleens of SNF_1 mice just before they begin to develop lupus nephritis and comprise two immunophenotypes, that is, $L3T4^+$, $Lyt-2^-$, $Thy-1^+$, Ig^- ($CD4^+/CD8^-$) and $L3T4^-$, $Lyt-1^+$, $Lyt-2^-$, $Thy-1^+$, Ig^- ($CD4^-/CD^-$). In addition, B cells capable of secreting the highly cationic anti-DNA antibodies are present only in older animals and are an expanded clone (Datta et al., 1987). IL-2-dependent $CD4^+/CD8^-$ and $CD4^-/CD8^-$ T-cell lines were derived from nephritic SNF_1 mice. While some cell lines of each T-cell phenotype can induce nephritogenic anti-DNA antibodies from SNF_1 B cells, the $CD4^+$ T-cell lines are highly autoreactive in culture. These $CD4^+$ cell lines could not induce cationic IgG anti-DNA antibodies from B cells derived from either the SWR or the NZB parents, suggesting that the parental strains may be deficient in select B cells committed to the production of nephritogenic anti-DNA antibodies (Sainis and Datta, 1988).

PN (Palmerston North)

Genetics

Inbred PN mice trace their ancestry to albino mice purchased from a pet store in New Zealand in 1948. From Palmerston North Hospital, Palmerston North, New Zealand, these "TW" outbred mice were sent to Massey University, then to Glaxo New Zealand Ltd., and eventually were returned to Palmerston North Hospital, where they were renamed PN. In 1964, under the direction of R. D. Wigley, inbreeding was begun. Selection for ANA positivity was carried out during the first three generations. The primary breeding line was named PN/n A. A second breeding line, PN/n B, was established after nine generations of inbreeding of PN/n A. PN/n A (at F28) and PN/n B (at F23) were imported into the United States in 1974 by S. E. Walker (subline code Sw). The PN/n A strain was arbitrarily discarded in 1975 (Walker et al., 1978).

The PN strain carries the MHC haplotype $H-2^q$. These mice are C-4 high, Slp negative, and G7 (C4d) negative (Schultz et al., 1982). In preliminary studies of inheritance of autoimmunity, PN/Sw mice crossed with the nonautoimmune DBA/2J strain produced offspring that did not develop proliferative glomerulonephritis or vasculitis. It was concluded that the autoimmune disease in PN mice has a recessive mode of inheritance.

Pathophysiology

PN mice are considered to be a model for SLE. The mean life span of female PN/Sw mice is 11.6 months. Males survive for an additional 4 months. Walker et al. (1978) determined that the most common causes of death are renal disease and vasculitis. Glomerulonephritis was present in 74 percent of autopsied mice. Mesangial thickening, hypercellularity, fibrinoid necrosis, and crescent formation were found. Perivascular infiltrates of lymphoid cells were found in nearly all kidneys of autopsied PN mice. Arteritis, especially prominent in lymphoid organs and spleen, was found in the majority of autopsied mice. A 13 percent incidence of lymphoma was described. In mice studied histologically from 1 to 21 months of age, hyperplastic lymph nodes were a common finding. Thymuses of mice 1 to 5 months of age were also hyperplastic.

PN/Sw × NZB/Sw F_1 or NZB/Sw × PN/Sw F_1 hybrid females develop anti-DNA antibodies and die prematurely (mean age, 43 weeks) with vasculitis, renal disease, and lymphomas. In contrast, hybrid males have divergent patterns of mortality. PN/Sw × NZB/Sw F_1 males have a mean longevity of 67 weeks; NZB/Sw × PN/Sw F_1 males have a mean longevity of 104 weeks (Walker, et al., 1986).

Unique among animal models of SLE is the finding of ANA or anti-DNA antibodies in some newborn PN/Sw mice. Anti-DNA antibodies are detected in serum of 51 percent of mice between 2 weeks and 2 months of age. At 1 year of age, all female and 78 percent of male PN/Sw mice have anti-DNA antibodies. Glomerular deposits of IgG1, IgM, IgA, and C3 are detected (Walker et al., 1978). All strains of autoimmune mice synthesize antihistone antibodies. In PN mice, the antihistone antibodies recognize preferentially H1 and H2B; this pattern of antihistone antibody reaction is seen in human SLE (Costa and Monier, 1986).

One interesting finding that differentiates the PN strain from other murine models of SLE is that their spleen and thymus cells do not express high levels of MuLV-related gp70 and cannot be induced to produce infectious ecotropic or xenotropic MuLVs (Davidson, 1982).

Walker and Hewett (1984) have described modest reductions in responsiveness of spleen cells to PHA and ConA in female, but not in male, PN/Sw mice after 24 weeks of age. Davidson (1982) has shown that 4- to 6-month-old PN mice show a marked increase in levels of serum IgG1, IgG2b, and IgA. PN mice also have an increase in spontaneous dinitrophenol (DNP)-specific PFCs but a decrease in PFC response to SRBCs. Even before severe disease symptoms are exhibited, the spleens of 2- to 3-month-old PN mice have a reduced frequency of detectable sIg-positive cells.

Studies of primary splenic PFC responses to thymus-dependent and thymus-independent antigens have shown that PN/Sw mice have defective in vivo IgG responses to thymus-dependent antigens. Defective production of IgG-specific plaques is evident as early as 3 weeks of age, is not influenced by aging to 43 weeks, and is not corrected by increasing the antigenic challenge 10-fold. Analyses of IgG subclasses show that there is a shift away from the expected predominance of IgG1-specific PFC (Walker, 1988).

Husbandry

Special husbandry procedures are not required.

Reproduction

These mice breed normally.

SAM-P (Senescence Accelerated Mouse)

Genetics

Inbred SAM strains were established from several pairs of AKR/J mice by continuous selection for short-lived litters. Two strains, SAM-P/1 and

SAM-P/2, were selected as being prone to early senescence. The SAM-P strains differ from AKR/J mice by two biochemical and two lymphocytic markers; however, both SAM-P and AKR/J are H-$2K^k$, I^k, D^k, suggesting that there was either an accidental crossing of AKR parents with other strains or that genetic polymorphisms were introduced into the SAM-P strains (Takeda et al., 1981).

Pathophysiology

SAM-P/1 and SAM-P/2 strains have a mean life span of 9 months when they are raised under conventional conditions (Hosokawa et al., 1987a). Postnatally they grow normally until about 6 months of age, at which time they develop a severe loss of physical activity, alopecia, coarse skin, periophthalmic lesions, lordokyphosis of the spine, cataracts, and osteoporosis. In one strain, SAM-P/1, spontaneous age-associated amyloidosis is characteristic, and a unique amyloid fibril protein has been isolated from the livers of SAM-P/1 mice (Higuchi et al., 1986).

Both SAM-P/1 and SAM-P/2 strains demonstrate markedly diminished antibody-forming capacity to T-cell-independent (DNP-Ficoll) and T-cell-dependent (SRBC) antigens compared with control AKR/J and C3H/He mice (Hosokawa et al., 1987a). In addition, NK cell activity also shows an early onset (2 months of age) of regression and a sharp decline from the level in control mice. In contrast, allospecific cytotoxic T-lymphocyte (CTL) responses from SAM-P strains remain at normal (control) levels until 6 months of age.

The cellular site of the defect in antibody response to T-cell-dependent (TD) antigen was investigated with a cell culture system. The Th-cell activity for the antibody response to TD antigen was impaired, but no evidence could be found for suppressor factors, abnormal adherent cells, or B cells (in the TD antigen assay). In contrast, the MLR and delayed-type hypersensitivity reactions were normal (Hosokawa et al., 1987b). These results suggest that two subsets of Th cells exist in SAM-P mice, one providing defective help for B-cell differentiation and the other providing normal activity for cell-mediated responses. It is still unclear whether the defect in T-cell function of SAM-P mice is controlled by a genetic defect expressed in the immune system or whether immunity is influenced by a genetic alteration in the microenvironment.

Husbandry

SAM-P strains develop clinical illness at 6 months of age and die by 9 months when they are maintained in a conventional facility. No attempts at barrier maintenance of SPF SAM-P mice have been reported.

Reproduction

No special breeding practices have been reported.

SJL/J

Genetics

The SJL/J strain was derived from noninbred Swiss Webster mice. Murphy (1963) described the appearance of lymphomas resembling those of Hodgkin's disease in this strain. Histologically, these lymphomas were classified as type B reticulum-cell sarcomas (RCSs) (Dunn, 1954; reviewed in Ponzio et al., 1986). The genetic basis for the development of RCSs in SJL/J mice is unknown. A single autosomal dominant gene, *Rcs-1*, that suppresses the appearance of spontaneous RCSs in SJL/J mice has been described in the A.SW strain, and *Rcs-1* has been shown to be distinct from *H-2* and from the genes affecting MuLV, that is, *Fv-4*, *Cv*, and *Fgv-1* (Bubbers, 1984). Neither the major histocompatibility locus, which is $H-2^s$ in SJL/J mice, nor the Igh-1 locus, which is $Igh-1^b$ in SJL/J mice, appears to affect the incidence of RCS (Bubbers, 1983; E. B. Jacobson, Merck Institute, Rahway, New Jersey, unpublished data). A study comparing the congenic strains SJL/J and SJL/J-*lpr*/+ has shown that the appearance of RCS is greatly accelerated in SJL/J-*lpr*/+ mice (Morse et al., 1985). While SJL/J-*bg/bg* mice have an incidence of RCS similar to that seen in the SJL/J strain, the appearance of these tumors is slightly accelerated in the mutant (J. B. Roths, The Jackson Laboratory, Bar Harbor, Maine, unpublished data).

SJL/J mice produce low levels of IgE because of a strong tendency to develop IgE-specific suppressor cells (Watanabe et al., 1976; Itaya and Ovary, 1979). This suppression is controlled by a single autosomal recessive gene that is not linked to either the *H-2* or the *Igh* locus.

Two genes have been described in SJL/J mice that mediate resistance to the demyelinating effects of the JHM strain of MHV (Stohlman et al., 1980). One of these genes is dominant (*Rhv-1*); the other is recessive (*rhv-2*). Neither of these is linked to the *H-2* or *Igh* locus (Stohlman and Frelinger, 1978). In contrast, SJL/J mice are much more susceptible than are mice of other inbred strains to demyelinating disease induced by Theiler's murine encephalomyelitis virus (Lipton and Dal Canto, 1979). The inheritance of this susceptibility is recessive and is dependent on the *H-2D* locus and one or more unlinked loci that contribute to susceptibility in a manner consistent with a gene dosage model (Lipton and Melvold, 1984; Clatch et al., 1985; Rodriguez et al., 1986).

SJL mice are highly susceptible to the induction of acute experimental allergic encephalomyelitis (EAE) by immunization with allogeneic spinal cord homogenate in adjuvant, and they later develop a relapsing form of EAE (Brown and McFarlin, 1981). Susceptibility to EAE is inherited as a dominant trait. One gene controlling sensitivity has been mapped to the *H-2K* region (Bernard, 1976). Other important genes regulating sensitivity to EAE are the recessive gene for vasoactive amine hypersensitivity and the dominant gene governing induction of histamine sensitivity by *Bordetella pertussis* (Linthicum and Frelinger, 1982). Both are expressed by SJL/J mice.

It has recently been discovered that in the SJL/J strain (Behlke et al., 1986), as well as in some known autoimmune strains (Singer et al., 1986), there is a deletion of a large portion of the V_β T-cell receptor genes, resulting in a skewed V_β gene repertoire. In addition, a subpopulation of T cells from SJL mice (approximately 10 percent of peripheral T cells) expresses the recently described V_β T-cell receptor allele $V_\beta 17a$, which is associated with responsiveness to I-E (class II MHC) surface antigens (Kappler et al., 1987). SJL/J mice have very low levels of circulating $\lambda 1$ immunoglobulin compared with other strains. This appears to be due to a transcriptional defect affecting the level of $\lambda 1$ synthesis, rather than to the numbers of B cells capable of $\lambda 1$ production (Sanchez et al., 1985).

Pathophysiology

At the age of 12.5–13.5 months, the incidence of RCS is approximately 90 percent in males and females, whether they are virgins or retired breeders (Murphy, 1979). Primary RCSs are thought to arise in germinal centers of Peyer's patches and mesenteric lymph nodes (Siegler and Rich, 1968) and to spread to other lymph nodes, spleen, liver, and ovaries (Haran-Ghera et al., 1973). The primary lesions contain mainly lymphocytes, as well as histiocytes, eosinophils, plasma cells, and Reed-Sternberg-like cells (McIntire and Law, 1967; Murphy, 1969; Kumar, 1983). RCSs are most likely of B-cell origin, because intravenously injected RCS cells exhibit typical B-cell homing patterns (Pattengale and Taylor, 1983), many of the cells found within primary lesions contain cytoplasmic immunoglobulin (Taylor, 1976), the surface properties of RCS cells are consistent with a B-cell derivation (Beisel and Lerman, 1981; Kincade et al., 1981; Scheid et al., 1981), and primary RCSs fail to arise in mice treated with goat antimouse µ-chain antiserum from birth (Katz et al., 1980).

The appearance of spontaneous RCSs is often accompanied by abnormalities in serum immunoglobulins. Paraproteins of the IgG1, IgG2a, and IgG2b isotypes are found in sera from SJL/J mice greater than 6 months of age (Wanebo et al., 1966). Although the appearance of these paraproteins is temporally correlated with the appearance of primary RCSs in SJL/J mice, it is unclear whether the paraproteins are a product of RCSs or are formed by the host in response to RCSs. It is possible that B-cell hyperstimulation in SJL/J mice is an important factor in the pathogenesis of both phenomena. They could also be two independent characteristics of these mice. Also related may be the high incidence of spontaneous amyloidosis in SJL mice (Scheinberg et al., 1976). The role of ecotropic MuLV in the etiology of spontaneous RCSs in SJL/J mice has not been clearly defined (DeRossi et al., 1983).

Spontaneous RCS development is a thymus-dependent process (Katz et al., 1981). This is of interest in light of the unusual thymic abnormalities in aging SJL/J mice. SJL/J mice have a secondary increase in thymic weight after 7 months of age that appears to be caused by an influx of surface immunoglobulin-positive B cells (Ben Yaakov and Haran-Ghera, 1975) that peak in SJL/J mice between 6 and 12 months of age prior to RCS development. Although the tendency toward B-cell proliferation in SJL/J mice might contribute to this penetration of B cells into the thymus, there is also a contribution of the thymic environment through a leaky blood–thymus barrier (Claësson et al., 1978).

Although SJL/J mice are sensitive to tolerance induction at birth, by 2 months of age, unlike other strains, they become quite resistant to tolerance induction by antigens such as aggregate-free rabbit gamma-globulin, human gamma-globulin, and bovine serum albumin (Fujiwara and Cinader, 1974a; Owens and Bonavida, 1976). The presence of endogenously activated macrophages (Crowle and May, 1978) seems to play an important role in this resistance to tolerance induction (Fujiwara and Cinader, 1974b). There is an age-dependent decline in thymic and peripheral suppressor T cells (Nakano and Cinader 1980; D. A. Clark et al., 1981) that also appears to contribute to the tolerance resistance and to a tendency to develop autoantibodies (Vladutiu and Rose, 1971; Bentwich et al., 1972; Bernard, 1976; Owens and Bonavida, 1976; Cooke and Hutchings, 1984). SJL/J mice are able to produce a Th-cell factor that acts on B cells in a T-cell-independent manner to promote immunoglobulin secretion earlier in life than in the A/J or C57BL/6J strains (Matsuzawa and Cinader, 1982). Production of this factor reaches a maximum by 10–12 weeks of age (as opposed to 20–30 weeks for the A/J and C57 BL/6J strains). Although the early onset of Th-cell factor production is correlated with the age-dependent decline in the generation of suppressor cells, attempts to establish a causal relationship between the two phenomena have failed (Amagai et al., 1982).

Ten- to twelve-month-old SJL/J mice (both normal and obviously tumor bearing) exhibit abnormalities in their abilities to mount graft-versus-host responses, to develop delayed-type hypersensitivity reactions, and to reject skin allografts (Haran-Ghera et al., 1973). Normal SJL/J mice have low levels of endogenous NK cell activity and are extremely resistant to NK cell induction by interferon (IFN), poly(I-C), and *Corynebacterium parvum* (Fitzgerald and Ponzio, 1981; Kaminsky et al., 1983). SJL/J-*nu/nu* mice have elevated levels of endogenous NK cell activity compared with those in

SJL/J-*nu*/+ or SJL/J +/+ mice, and their NK cells are responsive to IFN (Kaminsky et al., 1985).

It has been shown that resistance to central nervous system challenge with the JHM strain of MHV is a function of adherent cells, indicating that endogenously activated macrophages might be an important aspect of the resistance of SJL/J mice to the induction of demyelinating disease by the JHM strain of MHV (Stohlman et al., 1980).

Pathophysiologic aspects of the high sensitivity of the SJL/J mouse and some of its F1 hybrids to autoimmune diseases such as EAE include H-2^s-linked recognition by T cells of the autoantigen(s) in question. A second important factor is the tendency of mice such as SJL/J, which have a high vasoactive amine sensitivity and are suseptible to the *Bordetella pertussis*-induced increased histamine sensitivity, to develop leakiness of their blood–organ barriers (Teuscher, 1985). In the case of EAE, enhanced vascular permeability in brain tissue has been observed after *B. pertussis* administration to SJL/J × BALB/c F_1 hybrids, and the induction of EAE might be inhibited by vasoactive amine antagonists (Linthicum et al., 1982). Thus, SJL/J mice might be a good model for autoimmune disease, based on the strain's high lymphokine production and macrophage activity, decreased suppressor cell activity, high vasoactive amine sensitivity, and a skewed V_β T-cell receptor gene repertoire.

Husbandry

SJL/J and some of its F1 hybrid males become extremely aggressive after 8 weeks of age. Fighting is severe enough to require that older males be caged singly (Crispens, 1973). Alternatively, the nasal installation of zinc chloride solution is known to disrupt olfactory sensitivity to pheromones and has proven effective in preventing fighting among mice (Alberts, 1974). No other special procedures are necessary.

Reproduction

SJL/J mice breed well and have relatively large litters (average of 7.9 pups per litter) (Crispens, 1979); however, only about half survive to weaning.

SL/Ni

Genetics

The SL/Ni strain, which should not be confused with the *Sl* (Steel) mutation, was developed in 1970 by Y. Nishizuka (Aichi Cancer Center Research Institute, Nagoya, Japan) from the lymphoma-prone SL strain (Nishizuka et al., 1975). The SL/Ni strain exhibits a decreased incidence of lymphoma. Two related substrains are also available. One retains the high incidence of nonthymic lymphoma; the other possesses a dominant epistatic gene, *Slvr-1*, that selectively restricts the expression of endogenous ecotropic virus (Hiai et al., 1982).

Pathophysiology

SL/Ni mice show a high frequency of spontaneous arteritis and glomerulonephritis (Kyogoku, 1977, 1980; Nose et al., 1981). These lesions have histological features similar to those seen in other murine models of SLE, such as the NZB × NZW F_1 hybrid and the MRL/Mp inbred strain. The vascular lesions (segmental fibrinoid arteritis) histologically resemble those of human polyarteritis nodosa (Kyogoku et al., 1981). The most frequently affected vessels are medium- to small-sized arteries of the ovary, uterus, parotids, kidney, spleen, and pancreas (Miyazawa et al., 1987). Approximately 30–50 percent of SL/Ni mice manifest acute and chronic arteritis beginning at 9 months of age. The incidence is higher in females than in males, and nearly all multiparous females develop arteritis (Nishizuka, 1979). Most SL/Ni mice develop renal glomerular disease starting at 5–6 months of age (Yoshiki et al., 1979).

Granular deposition of gp70 immune complexes is found in vessel walls and along glomerular capillary loops and mesangium (Yoshiki et al., 1979; Miyazawa et al., 1987). The renal disease has a progressive and protracted clinical course and affects females earlier than males. C-type MuLV particles have been found in abundance around the smooth muscle cells of the vessels, and budding from the membranes of these cells has been observed (Nose et al., 1981; Miyazawa et al., 1987). A decrease in suppressor T-cell activity (Matsumoto, 1979), a reduced in vivo anti-SRBC PFC response (Matsumoto, 1979), and production of anti-gp70 antibodies (Miyazawa et al., 1987) are associated with the onset of disease. SL/Ni mice also produce anti-ssDNA autoantibodies resulting in large concentrations of circulating immune complexes (Kyogoku, 1980).

Husbandry

SL/Ni mice are extremely susceptible to Sendai virus infection. Breeding and maintenance in isolators or under SPF conditions is recommended.

Reproduction

SL/Ni mice are perpetuated by brother × sister mating. Females have a tendency to become infertile, because their vascular inflammatory lesions affect the uterine and ovarian arteries; therefore, they are bred as soon as they reach sexual maturity (M. Kyogoku, Department of Pathology, Tohoku University School of Medicine, Sendai, Japan, unpublished data).

OUTBRED MICE

SWAN (Swiss Antinuclear)

Genetics

The SWAN mouse stock was derived from a pair of Gif:S (Swiss) mice that were positive for ANA. Following an initial three generations of inbreeding, the SWAN stock was maintained by random matings as a closed colony (Monier and Robert, 1974).

Pathophysiology

The hallmark of SWAN mice is the development of ANA. Monier et al. (1971) determined that 100 percent of SWAN mice tested after 8 months of age were positive for ANA. Females are more precocious in the development of ANA than are males. Both neonatal thymectomy and treatment with Freund's adjuvant accelerate the age of onset of ANA positivity. SWAN mice are reported to have a decreased lymphoproliferative response to PHA, and the peak primary antibody response to SRBCs is delayed from the time of immunization and is reduced in magnitude (Monier and Robert, 1974). The studies of Blaineau et al. (1978) indicate that the autoimmune pathology of SWAN mice occurs in the absence of xenotropic-type C-virus production. SWAN mice, like NZB and BWF_1 hybrid mice, are reported to have a premature decline in secretion of circulating facteur thymique serique (FTS) (Bach et al., 1980).

SWAN mice have histopathologic features similar to those of human SLE. By 8 months of age, all SWAN mice have glomerulonephritis with immunoglobulin deposits in their glomeruli. Perivascular lymphoid infiltrates, mesangial thickening, hyalinization, thickened basement membrane, and dilation of the renal tubules are commonly observed. Amyloidosis is also frequently found (Monier et al., 1971). Histopathologic abnormalities at the dermoepidermal junction are also seen (Monier and Sepetjian, 1975). The life span of these mice is unknown.

Husbandry

Special husbandry procedures are not required.

Reproduction

SWAN mice are maintained by random breeding.

RAT MUTANTS

ia (Incisor Absent)

Genetics

Incisor absent (*ia*) is one of three autosomal recessive, osteopetrotic mutations in the rat (see also *op*, page 124, and *tl*, page 128). Although the chromosomal locations of these three mutations are not known, it has been shown that they are not allelic (Moutier et al., 1976). The mutation *ia* arose spontaneously in a stock of unknown genetic makeup (Greep, 1941). Older literature used *in* (incisorless) as the gene symbol (Castle and King, 1944).

Pathophysiology

Osteopetrotic rats are smaller than their normal littermates. Teeth fail to erupt and incisors are absent. Although the teeth do not erupt, they continue to grow. Consequently, large odontomatous masses develop in the jaws, frequently causing facial distortion (Schour et al., 1949; Marks, 1976a). Unlike the mouse, osteopetrotic mutations in the rat are not associated with pigmentation defects. In all three mutants, skeletal growth is abnormal; long bones have no flared ends or marrow cavities.

In *ia/ia* rats, bone resorption is reduced by about 30 percent compared to normal littermates (Marks, 1973). Homozygotes can be identified on the day after birth by radiographic examination, which reveals a clawlike appearance of the distal humerus and generalized radiopacity of the long bones (Marks, 1978). Marrow spaces begin to develop during the second postnatal week (Marks, 1976a) and approach normal proportions during the third month. However, the shape of the bones remains abnormal, and metaphyseal bone continues to be markedly sclerotic throughout life (Marks, 1976a, 1981). The skeletal defect in *ia/ia* rats appears to be cellular, not humoral (Nyberg and Marks, 1975).

Osteoclasts in *ia* homozygotes are more numerous than in their normal littermates (C. R. Marks et al., 1984; Miller and Marks, 1982), have elevated levels of acid phosphatase (Marks, 1973; Ek-Rylander et al., 1989), and lack cytoplasmic vacuoles and ruffled borders (Marks, 1973). Osteoclasts can produce enzymes for resorption but cannot deliver them efficiently to the bone surfaces because of the lack of ruffled borders.

Serum calcium levels in *ia* homozygotes are normal; however, there is a poor response to the administration of exogenous parathyroid hormone (Marks,

1973, 1977). Phosphate levels are reduced, at least up to 40 days postpartum (Kenny et al., 1958). Serum levels of 1,25 dihydroxyvitamin D are elevated, and levels of the 24,25 dihydroxy metabolite are reduced (Zerwekh et al., 1987).

The defect in *ia/ia* rats is curable by transplantation of bone marrow, thymus, or spleen cells from normal animals (Milhaud et al., 1975; Marks, 1976b, 1978; Marks and Schneider, 1978; Miller and Marks, 1982). Skeletal sclerosis can be induced by the transplantation of mutant spleen cells into immunosuppressed normal littermates (Marks, 1976b). The cell-mediated response of *ia* homozygotes to oxazolone stimulation is normal for both the skin and lymph node blastogenesis assays (Schneider, 1978).

Husbandry

The *ia* mutation is not lethal; however, soft diets are essential to compensate for the lack of incisors. The husbandry of this mutant has been reviewed (Marks, 1987). Homozygotes have a normal life span with delayed sexual maturity.

Reproduction

There is no evidence that the *ia* mutation impairs reproductivity; however, to ensure larger, better-cared-for litters, breeding heterozygous females with homozygous males is suggested.

op (Osteopetrosis)

Genetics

Osteopetrosis (*op*) is an autosomal recessive mutation that arose spontaneously in a random-bred stock carrying the mutation fatty (*fa*) (Moutier et al., 1973). It is not allelic to *ia* (page 123) and *tl* (page 128), the two other genes that confer osteopetrosis in the rat. The *op* gene is epistatic to *ia* (Moutier et al., 1974).

Pathophysiology

The *op* mutation is lethal; maximum life span has been reported to be 116 days (Milhaud et al., 1977). Skeletal manifestations are similar to those in *ia* homozygotes, except that there is no delayed development of marrow spaces. Bones contain numerous, large extracellular lipid inclusions. There are reduced numbers of osteoclasts (Marks and Popoff, in press), and those present exhibit pleomorphic shapes and have reduced levels of tartrate-resistant ATPase (Ek-Rylander et al., 1989). Therefore, resorption is reduced by a deficiency of both cells and enzymes. Bone resorption is not stimulated by transplantation of compatible normal thymic grafts, indicating that the defect lies in myeloid rather than in thymic regulation of osteoclast function (Nisbet et al., 1983). Serum levels of 1,25 dihydroxyvitamin D are elevated, and levels of the 24,24 dihydroxy metabolite are reduced (Zerwekh et al., 1987).

Milhaud et al. (1977) reported a progressive loss of responsiveness of thymic and splenic cells toward both T- and B-cell mitogens. Evans et al. (1985) could not confirm a decreased responsiveness for spleen cells; however, the response of thymic and lymph node T cells to PHA was lower in *op/op* rats over 12 days of age than in normal littermates of the same age. Osteopetrotic rats under 10 days of age did not show this deficit. Since osteopetrotic obliteration of the marrow cavities precedes the T-cell deficiency, the investigators suggest that this deficiency is secondary to the skeletal derangements.

Transplantation of bone marrow from normal littermates cures the disease (Milhaud et al., 1975). Adoptive transfer of the disease has not been reported.

Husbandry

Because they have no incisors, these animals must be fed soft diets. It is preferable that they be maintained in a pathogen-free environment. The husbandry of this mutant has been reviewed (Marks, 1987).

Reproduction

The *op* mutation is maintained by breeding heterozygotes, which reproduce satisfactorily. It is possible to breed homozygotes after thymus transplantation.

rnu (Rowett Nude); *rnu^N* (*nznu*, New Zealand Nude)

Genetics

Nude (*rnu*) is an autosomal recessive mutation that has not yet been mapped. It was first found in 1953 in an outbred colony of hooded rats at the Rowett Research Institute, Aberdeen, Scotland; however, its biomedical significance was not recognized. In 1977 athymic nude rats were again observed in this colony and were subsequently distributed (May et al., 1977). A second nude mutation in the rat arose in a colony of outbred albino rats at Victoria University, Wellington, New Zealand (Berridge et al., 1979) and was given the name New Zealand nude, *rnu^N* (synonyms, *nznu*, *rnu^{NZ}*). The two mutations are allelic (Festing et al., 1978), and breeding data indicate that *rnu* is dominant to rnu^N (H. J. Hedrich, Central Institute for Laboratory Animal Breeding, Hannover, Federal Republic of Germany, unpublished data). Both alleles have been backcrossed onto a variety of inbred strains, including DA, F344, LEW, PVG, and WAG.

Pathophysiology

Both *rnu/rnu* and rnu^N/rnu^N rats possess only rudimentary thymic tissue (Fossum et al., 1980; Vos et al., 1980), which becomes cystic during ontogeny, as in nude mice (Groscurth et al., 1974). There are, however, differences in immunologic responsiveness between the two mutations. Unfortunately, these differences have not been worked out in detail.

These mutants are also dissimilar with respect to the degree of hairlessness. Homozygous rnu^N rats have no hair; homozygous *rnu* rats have cyclic hair growth, sometimes producing a pelage that covers the entire body. The degree of hairlessness is not due to a genetic variation in the strain on which the genes are carried (Festing, 1981); the same phenotypic differences are present when *rnu* and rnu^N are backcrossed onto identical inbred strains.

Under conventional conditions, Rowett nude rats survive for approximately 9 months, whereas New Zealand nude rats rarely survive for longer than 4 months. Life expectancy appears to be genetically determined and not primarily a result of the microbiological status of the environment (H. J. Hedrich, Central Institute for Laboratory Animal Breeding, Hannover, Federal Republic of Germany, unpublished data).

In both nude mutations, lymphocytes are severely depleted in the thymus-dependent paracortical areas of all lymph nodes, and cortical germinal centers are sometimes absent. The thymus-dependent splenic periarteriolar sheaths are variably depleted of lymphocytes, and Peyer's patches are smaller than normal. Total leukocyte counts are nearly normal, lymphocyte counts are significantly lower, and neutrophil counts are elevated (Brooks et al., 1980; Fossum et al., 1980; Vos et al., 1980). In *rnu* homozygotes, serum immunoglobulins are nearly normal, although some variations are present: IgM, IgD, and IgG2b are normal; IgA, IgG1, and IgG2 are slightly elevated; and IgG2a is lower. Antigen retention by follicular dendritic cells ($OX2^+$) is similar in young homozygous *rnu/rnu* rats and heterozygous *rnu/+* rats passively infused with antibody. However, during active immunization, follicular dendritic cells of *rnu/rnu* rats fail to retain antigen, suggesting that antigen retention is dependent on specific antibody and not on the presence of thymus or functional T cells (Mjaaland and Fossum, 1987).

NK cell activity is markedly increased in these mutants, as it is in nude mice. B cells are relatively, but not absolutely, increased in lymphoid organs and the thoracic duct. Thymocytes and T lymphocytes are reduced in these sites, the most dramatic reduction being in the thoracic duct. Measurable numbers of T lymphocytes, identified by using the monoclonal antibody OX19, are present in the spleen, lymph nodes, and peripheral blood. Less than 2 percent of the cells are OX19$^+$ in *rnu/rnu* and *rnu^N/rnu^N* rats that are less than 8 weeks old. Whereas New Zealand nude rats show no increase in T-cell marker expression, Rowett nude rats older than 6 months of age have about 35 percent of lymph node cells that are OX19$^+$. No age-associated appearance of OX19$^+$ cells has been observed in heterozygous controls. Homozygous *rnu* rats are capable of responding to PHA or ConA, although the response is reduced. This responsiveness also is age dependent and can be altered by microbiological status. For example, spleen cells from germfree or SPF *rnu/rnu* rats (younger than 16 weeks of age) respond, whereas spleen cells from conventionally maintained animals are unresponsive (H. J. Hedrich, Central Institute for Laboratory Animal Breeding, Hannover, Federal Republic of Germany, unpublished data). Some *rnu/rnu* rats resist systemic and regional graft-versus-host disease (Rolstad et al., 1983). Mixed lymphocyte culture (MLC) reactivity is absent in young *rnu/rnu* rats but can be demonstrated in adult animals. Lymph node cells from *rnu/rnu* rats older than 6 months of age mount a clear-cut proliferative response against a stimulatory cell pool comprising different MHC haplotypes (Wonigeit et al., 1987). The responses, however, do not reach the levels observed in euthymic controls, which indicates a lower frequency of alloreactive cells in athymic animals. Measurable amounts of IL-2 are produced by these T cells, which indicates that the cells are most likely T helper cells. Furthermore, cytotoxic T-cell activity can be induced in *rnu/rnu* rats (Schwinzer et al., 1987). In addition, aged Rowett nude rats show allograft rejection in vivo. A random pattern of rejection can be observed, which indicates a marked difference in the individual clonal repertoire (Hedrich et al., 1987).

The reports on rejection or acceptance of normal xenogeneic skin are somewhat contradictory (Hedrich et al., 1987). An association between the success of xenogeneic tumor grafts and the age of the recipients has been clearly shown (Maruo et al., 1982); *rnu/rnu* rats should not be older than 8 weeks of age when grafted.

Because of the delayed and imperfect onset of T-cell function, these mutants are good models for investigating the thymus-independent maturation of T-cell precursors to functional effector cells in vitro as well as in vivo.

There is little information on the response of nude rats to typical rat pathogens. Unlike the nude mouse, nude rats "clear" experimental *Listeria monocytogenes* infection, although at a slower than normal rate. NK cell activity can be stimulated with bacille Calmette-Guérin (BCG) (an attenuated strain of *Mycobacterium bovis*) and *Corynebacterium parvum*, and this activity appears to be normal. Experimental Sendai virus infection causes respiratory distress, loss of bronchial epithelium, and interstitial pneumonia. Antigen persists for at least 32 days postinfection (Carthew and Sparrow, 1980; Sparrow, 1980); however, the ultimate outcome of the infection is unknown, and no circulating antibody has been detected. Chronic respiratory disease (*Mycoplasma pulmonis*), conjunctivitis, and periorbital abscesses have been reported as problems. Tyzzer's disease (*Bacillus piliformis*) can destroy a colony; young *rnu/rnu* rats are very susceptible and die within 2–3 weeks after exposure (Thunert et al., 1985). Thus, these animals can serve as sentinels. Although *Pneumocystis carinii* is not a specific pathogen for rats, nude rats are very susceptible to this parasite and spontaneously develop a severe pneumonitis when exposed to the organism (Ziefer et al., 1984).

Husbandry

Most husbandry practices include precautions similar to those used for nude mice, including use of cesarean derivation, barrier maintenance, limited access, filter hoods, laminar-flow devices, isolators, and autoclaved diet. These precautions appear to be warranted (Festing, 1981). Both nude mutations appear to be more sensitive to exogenous factors (microbial and physical environment) on an inbred than on an outbred background.

Reproduction

Both sexes of *rnu/rnu* rats are fertile, but homozygous females have difficulties in raising their young. The most profitable matings are usually between heterozygous (*rnu/+*) females and homozygous (*rnu/rnu*) nude males. Such matings produce 36 percent *rnu/rnu* offspring, which suggests that in utero loss occurs. Heterozygous matings produce the anticipated 25 percent homozygous nude pups. Under germfree and certain SPF conditions, homozygous (*rnu/rnu*) females litter approximately 0.16 young per female per week and raise approximately 0.05 young per female per week. New Zealand nude rats are more difficult to breed. Inbred background appears to affect the productivity of nude offspring.

tl (Toothless)

Genetics

Toothless (*tl*) is an autosomal recessive mutation that arose spontaneously in a partially inbred colony of Osborne-Mendel-derived rats (Cotton and Gaines, 1974).

Pathophysiology

Homozygous *tl* rats can be distinguished from their normal littermates by 10 days of age by their smaller size, short snout, and lack of incisors. Periorbital incrustation, possibly caused by defective lacrimal ducts, is common. Bones are thickened and lack marrow cavities. Osteoclasts are rare (Cotton and Gaines, 1974; Marks, 1977), and those present are small and have greatly reduced concentrations of acid hydrolases (Seifert et al., 1988). Toothless rats are hypophosphatemic and mildly hypocalcemic (Seifert et al., 1988) and respond poorly to exogenous parathyroid hormone (Marks, 1977).

Peritoneal macrophages in *tl/tl* rats are decreased 100-fold compared to normal littermates (Wiktor-Jedrzejczak et al., 1981). In vitro responsiveness of spleen cells to T- and B-cell mitogens exceeds that of normal littermates and appears to be a function primarily of the adherent cell population (Wiktor-Jedrzejczak et al., 1981). Toothless rats are not cured by transplantation of normal bone marrow or spleen cells (Marks, 1977), and skeletal sclerosis cannot be induced by transplantation of mutant spleen cells into normal animals (S. C. Marks, Jr., University of Massachusetts Medical School, unpublished data).

Husbandry

The *tl* mutation is not lethal; however, a soft diet is essential to compensate for the lack of incisors. The husbandry of this mutant has been reviewed (Marks, 1987). SPF conditions are recommended.

Reproduction

Homozygotes breed poorly or not at all. The mutation should be maintained by heterozygous matings.

C4 Deficiency

Genetics

A single-component deficiency for the fourth component of complement has been reported in Wistar rats (Arroyave et al., 1974). The mode of inheritance was reported to be autosomal recessive, with heterozygotes expressing 50 percent of normal levels of C4. The gene encoding for the complement fraction C4 has been mapped to the right of *RT1.B* in the rat MHC (Watters et al., 1987).

Pathophysiology

Arroyave et al. (1977) found that total hemolytic complement activity (CH_{50}) in these animals was 20 percent of normal. It could be restored to normal by the addition of purified human C4. No C4 inhibitor was found in the sera of affected rats that were capable of responding to the injection of normal rat serum by producing anti-C4 antibodies. Although both female and male rats were deficient, a significantly higher number of males were affected.

Husbandry

Special husbandry procedures are not required.

Reproduction

Although no precise mechanism has been identified, the reproductive efficiency of this colony is lower than expected (small litter size).

INBRED STRAINS OF RATS
BB/Wor and Other Sublines

Genetics

The BB [formerly called BB (BioBreeding) Wistar] rat develops an acute form of spontaneous juvenile insulin-dependent diabetes mellitus (IDDM) resembling human diabetes mellitus type I. The development of this model has been reviewed (Chappel and Chappel, 1983). The gene conferring susceptibility to diabetes has been shown to be associated with the MHC. Specifically, there is a requirement for the $RT1^u$ haplotype or a gene in close linkage with the gene coding for this haplotype (Colle et al., 1981). Non-MHC genes are also involved in the susceptibility to diabetes in this strain (Jackson et al., 1984).

Several features of the syndrome suggest the involvement of the class II u antigens ($RT1.B/D^u$) in the pathogenesis of IDDM (Goldner-Sauve et al., 1986). This is supported by the data of Buse et al. (1984), which are indicative of a restriction fragment length polymorphism (RFLP) in class II MHC genes. On the other hand, Kastern et al. (1984) have demonstrated by RFLP analysis that a BB/Wor-derived subline lacks a major proportion of class I MHC genes. Furthermore, Wonigeit has found that BB/Wor-derived rats express the *uv4* haplotype defined by a variation in atypical class I antigens ($RT1.C^{uv4}$) (K. Wonigeit, Klinik für Abdominal- und Transplantationschirurgie der Medizinischen Hochschule, Hannover, Federal Republic of Germany, personal communication to H. J. Hedrich, 1988).

The presence of the $RT1^u$ allele is not in itself sufficient to result in clinical manifestation of the disease. Additional factors are required for overt diabetes; a susceptibility for the development of insular, periductular, or intraacinar lymphocytic infiltration in the pancreas has been proposed. This susceptibility is thought to be coded for by the dominant gene *Pli* (pancreatic lymphocytic infiltration), which segregates independently of *RT1* (Colle et al. 1983).

Overt IDDM is strongly associated with a genetically controlled depression of circulating T lymphocytes (Guttmann et al., 1983). The nature of the T lymphopenia gene (l) is not yet known, but it has not been linked to the MHC. Whether the expression of the l gene is fortuitous or obligatory for IDDM needs to be verified. Herold et al. (in press) have reported on the results of a cross between an inbred (F25) diabetes-prone BB substrain with an incidence of IDDM of greater than 90 percent and an inbred (F25) diabetes-resistant BB substrain with no diabetes and no lymphopenia. The F1 offspring were normal, but in the F2 generation, the overall incidences of diabetes and lymphopenia were 30 and 27 percent, respectively. Lymphopenia was present in 76 percent of the diabetic rats but in only 5 percent of the nondiabetic animals. Furthermore, the diabetes occurred earlier in nonlymphopenic than in lymphopenic rats. Like et al. (1986b) have demonstrated that diabetes occurs among diabetes-resistant control lines (WA, WB, and WC) without lymphopenia.

Several groups are inbreeding BB rats originating from the outbred BioBreeding Wistar colony. Various lines exist with a variable degree of inbreeding and differences in the genetic background, except for the genes determining IDDM ($RT1^u$ and *Pli*, with or without l). Selective breeding has resulted in a 60–80 percent penetrance in affected substrains. Normoglycemic sublines (diabetes-resistant) with less than 1 percent incidence of IDDM are available. Recently, Guberski et al. (in press) demonstrated that diabetes-resistant BB/Wor rats have a gene(s) controlling the age of onset of diabetes. However, the genetic basis of the disease needs further investigation.

Pathophysiology

The disease is characterized by insulitis (mononuclear cell infiltration of the pancreas), which causes destruction of the islets of Langerhans (Seemayer et al., 1983). Acute clinical signs, including hyperglycemia (greater than 300 mg/dl), nonmeasurable insulin levels, ketoacidosis, hyperglucagonemia, polydipsia, and polyuria, develop between 6 weeks and 6 months of age (average age at manifestation is 3 months). The rats survive only if insulin is administered daily (Chappel and Chappel, 1983).

The characteristics of IDDM development in BB rats are suggestive of an autoimmune etiology. The development of overt hyperglycemia can be suppressed by immune modulatory techniques, including antilymphocyte serum, cyclosporine A, neonatal thymectomy, or injections of monoclonal antibodies specific to NK cells or CD4$^+$ T lymphocytes (Like et al., 1979, 1981, 1986a; Stiller et al., 1983). Furthermore, it is possible to prevent the manifestation of diabetes mellitus by the application of antibodies directed against class II MHC gene products (Boitard et al., 1985), by bone marrow transplantation (Naji et al., 1983), or by transfusion of whole blood from a nondiabetic subline without prior immunosuppression of the recipient (Rossini et al., 1985). The disease can be adoptively transferred to young, nonaffected, diabetes-prone rats or to immunosuppressed histocompatible rats by using ConA-activated spleen cells from acutely diabetic donors (Koevary et al., 1983).

In addition to anti-class II MHC antibodies, it has been shown that multiple autoantibodies react with parietal cells of the gastric mucosa, thyroid colloid, thyroid follicular cells, skeletal muscle, smooth muscle, and nuclear protein (Dyrberg et al., 1982; Elder et al., 1982; Baekkeskov et al., 1984). A Hashimoto-like thyroiditis has also been demonstrated (Sternthal et al., 1981). The presence of autoantibodies directed against lymphocytes is controversial (Dyrberg et al., 1982; Elder et al., 1982).

Diabetes-prone rats that develop diabetes mellitus always exhibit a severe impairment of the T-cell system. Both the helper/inducer and the cytotoxic/suppressor T-cell subsets in peripheral blood and lymphoid tissues are profoundly reduced (Bellgrau et al., 1982; Colle et al., 1983; Elder and MacLaren, 1983). An inversion of the ratio of helper/inducer T cells (CD4) to cytotoxic/suppressor T cells (CD8) occurs in all BB rats by about 60 days of age (Prud'homme et al., 1984). Woda et al. (1986) have shown that diabetes-prone BB rats lack the classic cytotoxic/suppressor T-lymphocyte subset (CD8). Furthermore, Greiner et al. (1986b) have demonstrated that RT7.1$^+$ T cells are depleted in diabetes-prone rats and that no RT6.1$^+$ T cells are detected in their peripheral tissues.

Functional analysis of lymphopenic animals shows a severe deficiency of all T-cell functions analyzed (Jackson et al., 1981; Elder and MacLaren, 1983), depressed skin allograft rejection across MHC and non-MHC barriers (Bellgrau et al., 1982; Klöting et al., 1984), defective proliferative responses to mitogens and to allogeneic cells in MLC (Rossini et al., 1983; Prud'homme et al., 1984), and defective cytotoxic T-cell function as measured by cell-mediated lympholysis (Prud'homme et al., 1988). The number of B cells and serum immunoglobulin levels are normal, and the number of PMNs is increased, apparently in response to various chronic respiratory infections. The in vitro analysis of T-cell functions suggests that most BB sublines exhibit immunologically incompetent T cells, which result from a postthymic or peripherally acquired maturational defect (Elder and MacLaren, 1983).

It is not clear whether the mechanism of β-cell destruction and its prevention are the same in lymphopenic and nonlymphopenic (less than 3 percent of diabetic-resistent lines) animals (Like et al., 1986b). Butler et al. (1988) suggest that a deficiency in suppressor cell activity results in an unchecked effector cell-induced β-cell cytotoxicity. NK cells or effector cells that do not display the classical cytotoxic/suppressor T-cell phenotype might also be involved (Greiner, 1986).

Guberski et al. (1988) were able to create an obese animal model of autoimmune diabetes mellitus by crossing Zucker female rats that were heterozygous for the gene fatty $(fa/+)$ to male diabetic BB/Wor $(+/+)$ rats. F1 hybrid females were backcrossed to BB/Wor diabetic males, and diabetic backcross males and females were mated. This latter mating fixed the recessive diabetes genes. Progeny from this mating were bred back to F1 hybrids, and progeny were selected that were genotypically diabetic and carriers of the fa gene. These rats served as parents for later inbreeding, which resulted in a new diabetic rat, BBZ/Wor. The BB/Wor diabetic rat resembles the lean BBZ/Wor rat $(fa/+$ or $+/+)$: both are nonobese; both have insulitis, comparable rates of diabetes, thyroiditis, similar islet cell pathology, and ketosis; and both require insulin therapy. Obese BBZ/Wor rats (fa/fa) are heavier and have prominent β-cell hyperplasia and degranulation. The obese BBZ/Wor rats are of the $RT1^u$ haplotype, are lymphopenic, and have an incidence of diabetes comparable to those of lean BBZ/Wor and BB/Wor rats. The obese BBZ/Wor rats, however, do not require exogenous insulin, most likely because of incomplete destruction of the pancreatic β cells (Guberski et al., 1988).

Husbandry

Maintaining BB rats is difficult because of their extreme susceptibility to opportunistic infections (Elder and MacLaren, 1983). It is therefore advisable to maintain these animals under laminar-flow, SPF, or gnotobiotic conditions. Gnotobiotic BB rats have the same incidence of diabetes as conventionally housed BB rats.

All animals must be weighed several times a week; when weight decreases or fails to increase, the urine should be tested for glucose and ketones using Tes-Tape® (Eli Lilly & Co., Indianapolis, Ind.) and Ketostix® (Miles Diagnostics, Elkhart, Ind.). When glucosuria is found, plasma glucose should be determined. Once glucose levels are above 250 mg/dl, insulin must be administered daily. The dose is determined by the amount of glucosuria and is usually between 1 and 2.5 units.

Reproduction

Diabetes in BB/Wor rats is associated with reduced fertility. In males, this appears to be associated with a primary disorder of Leydig's cells, which precedes changes in seminiferous tubules (Murray et al., 1983). Breeding efficiency in the high-incidence, diabetes-prone lines can be improved by administration of cyclosporine A, which delays the onset of the syndrome (Like et al., 1984), or whole-blood transfusions, which prevent the occurrence of hyperglycemia (Rossini et al., 1983).

Gross anatomical malformations observed in the offspring of diabetic BB rats are exencephaly, dysmaturity of the ossification process in cranial and long bones, and lumbosacral dysgenesis. Pups born to diabetic dams demonstrate significant delays in all aspects of motor development during the preweaning period. Control of maternal diabetes (regulation of blood glucose and prevention of ketosis) during pregnancy results in increased litter and fetal size, decreased perinatal mortality, and a significant reduction in the incidence of congenital malformation (Brownscheidle et al., 1983).

LOU/C

LOU/C rats spontaneously produce an IgG autoantibody that binds and neutralizes rat beta interferon (De Maeyer-Guignard et al., 1984). The titer of this autoantibody increases with age. This strain was selected for its high spontaneous incidence of myeloma, which is first seen at 8 months of age, and it has been postulated that the anti-IFN antibodies contribute to this disease (De Maeyer-Guignard et al., 1984). No studies have been conducted to investigate whether there is an increased susceptibility to infectious agents.

GUINEA PIG MUTANTS

C2 Deficiency

Genetics

Deficiency of the second component of complement (C2) in guinea pigs was the first C2 deficiency reported in an animal other than humans (Hammer et al., 1981). The gene for the defect behaves like a rare, silent allele of C2 that is MHC linked and inherited in the same way as normal C2 variants, that is, as an autosomal codominant trait. This mode of inheritance is similar to that in humans (Bitter-Suermann et al., 1981). The C2-deficient allele ($C2^0$) is linked to $C4^{s1}$ and Bf^F allotypes within the guinea pig MHC.

Pathophysiology

C2 deficiency is characterized by a total absence of the second component of complement, as measured by hemolytic C2 and antigenic C2 protein assays. It is the most frequently seen complement deficiency in humans, with an incidence approaching 1 percent (Peltier, 1982). Macrophages in C2-deficient guinea pigs synthesize an abnormal C2-like protein rather than functionally active C2 (Goldberger et al., 1982).

C2-deficient guinea pigs show no unique susceptibility to infectious diseases. Changes in the vascular permeability of skin induced by injection of the C1s subcomponent of the first component of complement are absent in C2-deficient guinea pigs but are present in C4-deficient guinea pigs, which suggests that the permeability agent is derived from C2 itself (Strang et al., 1986). The C2-deficient guinea pig is unable to make a good antibody to the T-cell-dependent antigen, bacteriophage ΦX174. The antibody responses are characterized by low titers, failure to detect amplification of the secondary response, and no IgM to IgG switch (Ochs et al., 1986). Humans with C2 deficiency have a high incidence of SLE, discoid lupus, and Henoch-Schönlein purpura. The C2 deficiency is more common in women than in men, and the incidence of lupuslike syndromes is more common in C2-deficient women than in C2-deficient men. C2-deficient guinea pigs have hypergammaglobulinemia, anti-DNP antibodies, and rheumatoid factor; however, overt autoimmune disease is not seen (Böttger et al., 1986a). This guinea pig model might be a good model for studying the relationship between C2 deficiency and SLE in humans.

Husbandry

Special husbandry procedures are not required.

Reproduction

These animals breed normally.

C3 Deficiency

Genetics

The defect in the third component of complement (C3) arose spontaneously in strain 2 guinea pigs, was expressed in an autosomal codominant fashion, and was not linked to the MHC or to the gene controlling expression of the C3a receptor (Burger et al., 1986).

Pathophysiology

Homozygous C3-deficient (C3D) guinea pigs have a functional C3 titer and antigenic activity that is only 6 percent of normal. Serum from these animals has a markedly reduced hemolytic complement and bacteriocidal activity in vitro, and homozygous deficient animals have impaired antibody synthesis to the T-cell-dependent antigen, bacteriophage ΦX174 (Böttger et al., 1986b). In addition, C3D guinea pigs have a defect in isotype switching from IgM to IgG. These in vivo abnormalities in humoral immunity are shared with guinea pigs with deficiencies of C2 or C4 (Böttger et al., 1985). However, in contrast to C2D and C4D guinea pigs, C3D guinea pigs do not have elevated IgM levels, nor do they have circulating immune complexes (Böttger et al., 1986b). Macrophages and hepatocytes from guinea pigs with homozygous C3D synthesize and secrete C3 in normal amounts. When analyzed by sodium dodecyl sulfate (SDS) polyacrylamide-gel electrophoresis and immunoblotting, the product of C3D hepatocytes appeared normal; however, it was greatly reduced in the serum of animals containing the deficiency. The catabolism of normal guinea pig C3 was not elevated in C3D guinea pigs; however, enhanced proteolysis of a defective C3 molecule in C3D animals could not be excluded (Auerbach et al., 1985).

Husbandry

C3D guinea pigs, unlike humans with a similar deficiency, do not appear to have an increased susceptibility to infectious agents or immune complex disease (Alper and Rosen, 1984; Böttger et al., 1986b).

Reproduction

No problems have been reported in the breeding of C3D guinea pigs.

C4 Deficiency

Genetics

Deficiency of the fourth component of complement (C4) in guinea pigs is inherited as an autosomal recessive trait with full penetrance (Hyde, 1923; Ellman et al., 1970), resembling the inheritance of C4 deficiency in humans (Ochs et al., 1977; Schaller et al., 1977). Heterozygotes express intermediate levels of C4 (Ellman et al., 1970). It is thought that the deficiency arose from a mutation of a *C4-F* allele (Shevach et al., 1976).

Pathophysiology

C4 deficiency was first discovered in a strain of guinea pigs whose serum did not lyse antibody-sensitized horse erythrocytes (Moore, 1919). The defect was discovered independently in a colony of outbred Hartley guinea pigs at the National Institutes of Health (Ellman et al., 1970). Guinea pigs with a C4 deficiency are unable to activate complement by the classical pathway (Hammer et al., 1981), because C4 is critical for the development of both C3 and C5 convertase (Müller-Eberhard, 1975).

This model has been used extensively to demonstrate the amino acid sequence of pro-C4 (the precursor to C4). The structure, function, and quantitation of guinea pig C4 have recently been reviewed (Quimby and Dillingham, 1988). The discovery of C4 deficiency in guinea pigs provided a good model in which to study the activities of the classical and alternative pathways of complement activation (Frank et al., 1971).

The genetic control and biosynthesis of C4 have also been studied in the C4-deficient guinea pig (Colten and Frank, 1972; Colten, 1983). Recent experiments with cell-free biosynthetic systems have shown that guinea pig C4 is synthesized as the single-stranded precursor pro-C4 (200,000 dalton), which, prior to its release from the polysome, is converted to the three-chain C4 molecule (Hall and Colten, 1977). Hall and Colten (1977) failed to detect the presence of pro-C4 in hepatocytes of C4-deficient guinea pigs, and Colten (1983) has hypothesized, based on hybridoma studies, that the deficiency is due to a posttranscriptional defect in the processing of C4 precursor RNA to mature C4 messenger RNA.

The C4-deficient guinea pig has also been used to evaluate the role of the classical pathway during various infections (Diamond et al., 1974; Gelfand et al., 1978). Early studies showed that animals with suspected C4 deficiency were more susceptible to *Salmonella* (formerly *Bacillus*) *choleraesuis* infection than were normal animals (Moore, 1919). C4-deficient guinea pigs are more susceptible to the lethal effects of endotoxin injection (May et al., 1972). Resistance to the ixodid tick *Dermacentor andersoni* (Wikel, 1979) and to *Candida albicans* (Gelfand et al., 1978) and *Cryptococcus neoformans* (Diamond et al., 1974) infections is normal in C4-deficient animals. It can be inferred, therefore, that there is an intact alternative pathway of complement activation in guinea pigs with these infections.

Ellman et al. (1971) have demonstrated a slight but significant defect in the ability of C4-deficient guinea pigs to make an antibody response to DNP–bovine γ-globulin. The antibody response to the T-cell-dependent antigen bacteriophage ΦX174 is clearly abnormal in C4-deficient guinea pigs. Not only is the IgM response depressed, but there is also no evidence for immunologic memory (Ochs et al., 1978, 1986). C4-deficient guinea pigs have signs of polyclonally stimulated antibody synthesis, circulating rheumatoid factors, and anti-DNP antibodies, which suggests immune complex disease (Böttger et al., 1986a). Despite the latter observation, there have not been any particular problems associated with infectious agents in conventionally reared colonies (Peltier, 1982).

Husbandry

Special husbandry procedures are not required.

Reproduction

These animals breed normally (Peltier, 1982).

HAMSTER MUTANTS

nu **(Nude)**

Genetics

Nude (*nu*) is a recessive mutation that arose spontaneously in the breeding colony of outbred Syrian hamsters at the Institut de Recherches Scientifiques sur le Cancer, Villejuif, France (Haddada et al., 1982). Genetic studies involving these animals have not been conducted.

Pathophysiology

Nude hamsters, like nude mice and nude rats, are hairless and have only a rudimentary thymus. At 1 month of age, decreased numbers of mature lymphocytes are found in lymph nodes and the spleen. Spleen cells do not proliferate in response to the mitogens ConA, PHA, LPS, and protein A, and the level of serum IgG is reduced. In contrast to nude mice and nude rats, nude hamsters have no higher NK cell activity than do normal hamsters. Spontaneous tumors do not arise frequently; however, a study has shown that 1 of 11 animals developed an immunoblastic sarcoma in the mesenteric lymph node at 8 months of age (Loridon-Rosa et al., 1988). Injection of simian virus 40-transformed B-lymphoma cells induces tumor development in nude hamsters but not in immunocompetent controls (Loridon-Rosa et al., 1988).

Husbandry

No definitive studies have been performed to evaluate the susceptibility of nude hamsters to various infectious agents; however, animals housed under conventional conditions live 10–12 months and often develop skin disorders after 8 months. Because the immunologic defect in nude hamsters appears to be similar to that of nude mice and rats, SPF conditions are recommended for their maintenance.

Reproduction

Female homozygous nude hamsters are fertile but are unable to feed their young; therefore, nude hamsters are maintained by mating homozygous nude males with heterozygous female siblings.

C6 Deficiency

Genetics

The genetic control of the deficiency of the sixth component of complement (C6) in hamsters is incompletely understood.

Pathophysiology

C6 deficiency in hamsters was described by Yang and coworkers (1974). Serum from affected hamsters was incapable of reconstituting C6-deficient rabbit or human sera but did reconstitute sera depleted of C1r, C2, and C4. The C6 deficiency was confirmed by titration of isolated complement components from normal and deficient hamster sera. C6-deficient hamster sera did not have altered immune adherence or phagocytic functions. Hamsters with the deficiency had a high incidence of proliferative enteritis (transmissible ileal hyperplasia).

Husbandry

Special husbandry procedures are not required.

Reproduction

These animals breed normally.

3

Induced Immunodeficiencies

Small rodents are used extensively to study experimentally induced immunosuppression and, occasionally, through a variety of circumstances, are also the unintended objects of induced immunosuppression. The following discussion will provide the reader with examples of agents that advertently or inadvertently induce suppression of immune responses in rodents. A complete review of this subject can be found in *Immunotoxicology of Drugs and Chemicals*, 2d ed. (Descotes, 1988).

CHEMICAL INDUCERS

A wide variety of chemicals and drugs have been used, particularly in the mouse, to study and measure immune function and suppressive effects. Tetrachlorodibenza-p-dioxin (TCDD) and polybrominated biphenyls (PBBs), for example, have been shown to generally impair immune responsiveness, and Tris depresses cell-mediated immunity (Luster et al., 1987).

Diethylstilbestrol

The immunosuppressive effects of diethylstilbestrol (DES) and a variety of other nonsteroidal and related estrogenic compounds have been studied in detail. DES is a recognized stimulant of the reticuloendothelial system (RES) in adult mice; however, when female mice are exposed to DES in utero or postnatally, thymic involution is induced (Luster et al., 1979). In addition, both T-cell-dependent and T-cell-independent antibody responses are suppressed, hypersensitivity responses are delayed, lymphoproliferative responses to T-cell mitogens are increased, and the numbers of hematopoietic stem cells and granulocyte-macrophage precursors in the bone marrow are depressed. Male mice exposed to DES in utero respond with an enhanced T-cell-dependent antibody response.

Heavy Metals

Heavy metals such as lead and cadmium can impair immune responsiveness. Although accidental exposure to toxic amounts of heavy metals in the laboratory environment would be unusual, the investigator should be aware of the influence that heavy metals can have on the immune system.

Lead

Lead has been shown to increase the susceptibility of rats and mice to bacterial, viral, and protozoal challenge (Koller, 1981). Lead-treated animals express decreased PMN phagocytic activity, and immunoglobulin and complement responses are reduced when lead-treated animals are subjected to active immunization regimens. Lead appears to block complement receptors on B lymphocytes (Koller and Brauner, 1977); however, lead does not significantly alter MLC responsiveness (Koller and Roan, 1980) or lymphocyte transformation by ConA and LPS (Koller et al., 1979). Lead also inhibits mitogenic responsiveness to PHA and pokeweed mitogen (PWM) and impairs the delayed-type hypersensitivity response.

Cadmium

Acute exposure to cadmium appears to produce mixed immunoresponsiveness in mice and rats; however, chronic exposure results in a suppression of antibody responsiveness that lasts for several weeks after exposure ceases (Koller et al., 1975). IgG is most affected, but Th cells can also be involved. E-rosette formation of alveolar macrophages and cytotoxicity of phagocytic cells have also been demonstrated (Hadley et al., 1977; Loose et al., 1978b). Cell-mediated immune responses to cadmium are mixed: The LPS response is enhanced, PHA and PWM responses are suppressed, and the MLC response is not affected (Koller, 1981).

Phenols

Archer et al. (1981) have summarized the effects of selected phenols on the immune system of mice. Butylatedhydroxyanisole (BHA), gallic acid (GA), and propylgallate (PG) suppress the primary anti-SRBC PFC response of mouse spleen cells. GA and PG also reduce the secondary response, but BHA does not. BHA, but not GA, suppresses the thymus-independent antibody response of nude mice.

Alkylating Agents

Alkylating agents are potent immunosuppressants in animals and humans. It is unlikely that accidental exposure would occur in experimental animals; however, these agents have been used extensively to study immune system function, particularly in mice. Alkylating agents suppress primary T-cell-dependent antibody responses when they are given prior to antigenic exposure (Ghaffer et al., 1981) or shortly after priming. They do not affect secondary responses unless the agent is administered after exposure to the antigen.

Glucocorticoids

Glucocorticoids are potent immunosuppressants in small rodents and are sometimes used as diagnostic aids to potentiate the expression of subclinical infections such as mouse hepatitis virus (MHV) and *Pneumocystis carinii*.

An immunoinhibitory role of glucocorticoids has been suggested from the observations that in rodents, thymic atrophy results from the administration of exogenous glucocorticoids or from prolonged stress (Stein, 1985). Although the effect of glucocorticoids on murine and guinea pig lymphocytes is much greater than that seen in humans, there is ample evidence that glucocorticoids have some inhibitory activity in all mammalian species studied (Claman, 1972; Calvano, 1986).

In rodents, the administration of glucocorticoids appears to have opposing effects, resulting in both the enhancement of antibody production (Bradley and Mishell, 1981) and T-cell depletion (Durant, 1986). Recently, in vitro studies have demonstrated that human $CD8^+$ T cells are more sensitive than Th cells to the effects of glucocorticoids (Paavonen, 1985). Other immunologic changes resulting from the administration of glucocorticoids include a depression in NK cell function (Fernandes et al., 1975), T-cell blastogenesis (Blomgren and Andersson, 1976), graft-versus-host reactions (Medawar and Sparrow, 1956), and delayed-type hypersensitivity (Gabrielsen and Good, 1967).

It has also been postulated that glucocorticoids affect immune cells by modulating the activity of macrophages. Warren and Vogel (1985) have shown that the gamma-interferon-induced increase in murine macrophage Fc-mediated phagocytosis is enhanced by glucocorticoids, while the levels of Ia antigen are decreased. Besedovsky et al. (1985) have provided evidence that stimulated rat mononuclear cells produce the hormone glucocorticoid increasing factor (GIF), which increases adrenal glucocorticoid output by a mechanism that requires the pituitary gland. One candidate for this GIF activity is IL-1, which has been shown to possess corticotropin-releasing activity (Woloski et al., 1985).

Exogenously administered glucocorticoids can profoundly affect the susceptibility of rodents to several infectious agents (Robinson et al., 1974). Glucocorticoids compromise the ability of rodents to combat infections caused by *Salmonella typhimurium* and *Babesia rodhaini* (Plant et al., 1983; Zivkovic et al., 1985) and make them susceptible to clinical illness caused by *Pneumocystis carinii* and *Bacillis piliformis*, which generally are not associated with overt disease (Hughes et al., 1983). A review of the action of glucocorticoids, cyclophosphamide, and other immunosuppressive drugs has been published recently (Sternberg and Parker, 1988).

INFECTIOUS AGENTS

Infectious agents and their products can be potent modifiers of the immune response through a number of mechanisms, including antigenic competition, preemption of antigen-reactive cells, activation of suppressor cells and generation of suppressive factors, lymphocytotoxicity, RES blockade, and alterations in lymphocyte interactions (Weidanz et al., 1978).

Small rodents have been used extensively to study many aspects of infection-regulated immunosuppression and are frequently victims of adventitious infections in the experimental laboratory. Considerable variation in response can be expressed by different rodent strains to such agents as Sendai virus (Parker et al., 1978), ectromelia virus (Bhatt and Jacoby, 1986; Wallace and Buller, 1986), and MHV (Levy et al., 1981); however, specific immunologic perturbations caused by these agents are not always well identified. Levy et al. (1981) have related MHV susceptibility directly to an increased coagulation protease activity by host monocytes following T-cell instruction. Boorman et al. (1982) have demonstrated peritoneal macrophage activation in C57BL/6 × C3H F_1 hybrids accidentally infected with MHV. These macrophages expressed increased tumor cell killing activity and had highly convoluted cell membranes. Mouse thymic virus, an uncommon herpesvirus that attacks the thymus gland of neonates causing severe thymocyte necrosis, diminishes the reactivity of spleen cells to T-cell mitogens and allogeneic cells. Thymocytes express a marked reduction in graft-versus-host reactivity during a period of several weeks after infection. Spleen cells are similarly affected, but lymph node cells are not (Cross et al., 1976).

Conversely, induced immunosuppression can lead to activation of latent or persistent murine infections such as parvoviruses (Tattersall and Cotmore, 1986), K virus (Greenlee, 1981), mouse adenovirus (Hashimoto et al., 1973), cytomegalovirus (CMV) (Osborn, 1986), and others that interfere with experimental results by causing mortality or shifts in baseline values. Elimination of adventitious infections by using sound colony management procedures and reliable suppliers is considered to be an important method for refining animal experimentation.

NUTRITION

Malnutrition is associated with atrophy of the thymus, Peyer's patches, Hassall's corpuscles, and T-cell-dependent pericortical and periarterial areas of lymph nodes and the spleen, respectively. There is also a reduction in primary follicles and in delayed-type cutaneous hypersensitivity. Humoral responses are less affected. Chronic restriction of protein results in depressed production of antibodies, and severe protein restriction depresses both humoral and cellular immunity (Jose and Good, 1973a,b). Calorie restriction can also reduce humoral response and increase T-cell suppressor activity (Fernandes et al., 1975; Fernandes, 1984). An increase in dietary polyunsaturated fatty acids (PUFAs) inhibits the cell-mediated immune (CMI) response and cytotoxicity in the in vitro ^{51}Cr release assay (Mertin, 1976). Protein calorie malnutrition can reduce antibody affinity for antigens (Katz, 1978). The lymphocytes of malnourished individuals have a reduced in vitro response to PHA stimulation. This might reflect deficient T-cell function or reduced numbers of T cells.

Dietary deficiencies of certain trace elements can have a profound suppressive effect on immunity in rodents. Rats fed magnesium-deficient diets develop thymomas, leading to an immunodeficient state (Bois et al., 1969; Hass et al., 1981; Averdunk et al., 1982). Moreover, magnesium-deficient diets lead to a reduction of serum immunoglobulins in the rat (Alcock and Shils, 1974; Rayssiguier et al., 1977). Deficiency of dietary zinc results in thymic atrophy and reduced Th cell function in mice (Fraker et al., 1977).

Inadvertent general malnutrition in laboratory rodents is rare, provided that standard commercial diets are fed. Impaired immunity caused by the accidental exclusion of specific nutrients also appears to be rare. The role of diet on immune function in rodents has been reviewed recently (ICLAS, 1987).

IONIZING AND ULTRAVIOLET RADIATION

Ionizing radiation serves as the classic example of a method for experimentally inducing immunosuppression in rodents, particularly the mouse. Radiation biologists who used this technique soon found that lethally or sublethally irradiated mice died more quickly than expected if *Pseudomonas aeruginosa* was present in the oral cavity or gastrointestinal tract. Furthermore, mice maintained under conventional conditions exhibited a wide variation in morbidity and mortality when they were exposed to total lymphoid irradiation (TLI) with specific shielding to protect bone marrow. This was apparently due to the presence in the holding rooms of endemic viral and bacterial organisms. Following TLI, B-lymphocyte numbers begin to rise in about 2 weeks; however, T lymphocytes remain completely suppressed for about 2 months, and T lymphocytopenia persists in the mouse for about 1 year (Slavin et al., 1977).

TLI modifies the MLR by heightening the response of peripheral blood lymphocytes (PBLs) to allogeneic lymphocytes while suppressing the response to syngeneic lymphocytes. The response of PBLs to ConA is initially suppressed but soon exceeds control values; the response to PHA is impaired for prolonged periods (Strober et al., 1979). Although most subpopulations of spleen cells are at normal values by 1 month after TLI, TL^+ (immature T) cells in spleen and lymph node populations are elevated. The reason for this is not entirely clear. Allograft survival approaches five times the control value following TLI, is about two times the control value if the thymus is shielded during irradiation, and is lower still if only the thymus is irradiated (Slavin et al., 1977). Antibody response to SRBCs is eliminated for about 1 month following TLI. Subsequently, IgM is produced, but IgG is not seen until about 7 months following TLI, and then it is produced at reduced levels. Irradiation of the thymus or subdiaphragmatic tissue has only a small effect on the response to SRBCs (Strober et al., 1979).

Immunosuppression caused by ionizing radiation is a function of cell population sensitivity to the killing effect of the radiation. Among the most sensitive populations are stem cells. Small lymphocytes are also highly radiosensitive (Casarett, 1980). Thus, ionizing radiation-induced immunosuppression is broad based and profound at relatively modest whole-body dose levels.

Exposure to ultraviolet radiation (UVR) also results in modifications to immunologic response. UVR reduces the density of Langerhans' cells expressing Ia and membrane-bound adenosine triphosphatase (ATPase), interferes with the antigen-presenting capacity of epidermal cells, causes an unresponsiveness to contact allergens, and interferes with the rejection of highly antigenic UVR-induced tumors (reviewed by Breathnach and Katz, 1986).

Little is known about microwave irradiation, although some of its nonthermal effects on antibody production and numbers of antibody-producing cells have been reported (Czerski et al., 1974).

BIOLOGICAL INDUCERS

Biological materials are commonly used as immunosuppressants in laboratory rodents. Among the most potent are anti-lymphocyte serum (ALS) and anti-theta serum, with or without added complement. Interferon poly(A-

U), transplanted tumors, and stress are also potent immunosuppressants (Chen and Goldstein, 1985; Johnson, 1985). Immunologic tolerance to specific antigens can be induced in neonatal rats and mice by transferring syngeneic spleen cells coupled with a palmitoyl derivative of protein antigens. This produces carrier-specific tolerance in mice, which is thought to be caused by the induction of T-cell tolerance and T-cell suppression and appears to decrease the T-cell help needed by specific B cells (Sherr et al., 1979). Glucan [$\alpha\beta$-(1,3-glucosidic polyglucose)] is a potent RES stimulant; however, it depresses NK cell activity in mice and might promote tumor growth (Lotzova and Gutterman, 1979).

Neoplasms can cause immunosuppression by invading and destroying normal immune system tissues, by serving as functional modulators of immune capacity (e.g., plasmacytomas or type-specific lymphocytomas), or by promoting uneven stimulation of immune modulators. The effects of stimulating the activity of intrinsic factors such as lymphokines, monokines, and thymic hormone are not always clear-cut; such stimulation might lead to immunosuppression (e.g., the possible promotion of T suppressor cells by thymosin and thymopoietin) (Chen and Goldstein, 1985).

THYMECTOMY

Surgical extirpation has long been used as an experimental method for ablating or modulating the immune response of small rodents. Thymectomy, including neonatal thymectomy, is a valuable experimental tool for immunologic investigation. Surgical ablation of the thymus gland has been shown to compromise immune function in mammals and birds (Miller et al., 1962). The procedure, which has been described for both rodents and birds by Hudson and Hay (1976), must be performed within 24 hours of birth in mice for maximal T-cell depression. Neonatally thymectomized mice that are not maintained under germfree conditions develop a wasting syndrome between 1 and 3 months of age, which possibly results from their inability to combat various infectious agents (Miller et al., 1962). Male mice thymectomized on the third postnatal day develop spontaneous autoimmune prostatitis at puberty by autosensitization to antigens that are normally expressed during prostate differentiation (Taguchi et al., 1985).

The study of neonatally thymectomized mice has been instrumental in delineating the T- and B-cell systems. Miller and Mitchell (1969) demonstrated that there is a reduced number of splenic PFCs when thymectomized mice are challenged with SRBCs. Similarly, Davies (1969) demonstrated a decreased response of humoral antibodies of the IgG (mercaptoethanol-resistant) class in thymectomized mice injected with horse erythrocytes. Neonatal thymectomy has been associated with a diminished humoral response to a vaccine strain of Japanese B encephalitis virus (Mori et al., 1970) and herpes simplex virus (A. C. Allison, Clinical Research Centre, Harrow, Middlesex, England, unpublished data).

The ability of the thymectomized host to combat infections has also received attention. Thymectomized mice inoculated with ectromelia virus developed a fatal, generalized infection (Allison, 1974). Likewise, thymectomized mice inoculated with herpes simplex virus type 1 developed fatal encephalitis (Mori et al., 1967). It is of interest, however, that thymectomized mice did not die following experimental infection with lymphocytic choriomeningitis virus (Levey et al., 1963), which is known to be mediated by T cells. There is a decreased ability to expel *Trichinella spiralis* because there is no increase in intestinal mast cell numbers associated with the parasitism in neonatally thymectomized mice (Brown et al., 1981). In this model the normal function of intestinal mast cells is dependent on the presence of thymic-derived lymphocytes.

Development of the nude mouse, and more recently the nude rat, has diminished the need for thymectomy procedures. Spleen cells from thymectomized mice express increased lytic activity against syngeneic neoplasms in vitro. Thymectomized polyomavirus-infected mice have been shown to have higher neoplasm rates than normal controls.

Splenectomy and lymphadenectomy are also used as immune system modulators in small rodents, although the latter is less commonly used.

4
Maintenance of Rodents Requiring Isolation

GENERAL CONSIDERATIONS

Certain mutations (i.e., nude mice, rats, hamsters; *scid* mice; and all multiple mutant strains carrying the homozygous nude gene) and experimental manipulations (e.g., sublethal irradiation) result in an immunodeficiency that is so severe that exposure of these animals to agents of even relatively low pathogenicity results in severe illness and death. For example, several studies have shown that athymic mice exposed to mouse hepatitis and Sendai viruses have survival times of less than 3 weeks (Sebesteny and Hill, 1974; Ward et al., 1976). However, in animal facilities that exclude pathogens to which they are susceptible, athymic mice exhibit a life span nearly as long as that of their euthymic counterparts (ILAR, 1976). It is therefore essential that the housing provided for these animals creates a barrier between the susceptible host and infectious organisms.

The *Guide for the Care and Use of Laboratory Animals* (NRC, 1985) states: "The caging or housing system is one of the most important elements in the physical and social environment of research animals" (p. 11). However, equally important in the maintenance of a barrier are proper husbandry procedures, personnel management, and equipment maintenance. To ensure the health of pathogen-free animals, programs should be designed to monitor the colony by using clinical, serologic, pathologic, microbiologic, and parasitologic techniques (Hsu et al., 1980; Fox et al., 1984; Small, 1984). Because the well-being of pathogen-free animals is so strongly influenced by the housing system, special design and environmental control are of utmost concern. These systems are described below.

SPECIAL FACILITIES AND EQUIPMENT

Plastic Cages with Filter Covers

This system consists of transparent plastic cages covered with securely fitted filter caps or covers. All components can be sterilized separately by autoclaving. The committee recommends the use of transparent plastic cages to facilitate routine animal observation without the need to open the cage more frequently than is necessary for sanitation and experimentation. These units represent the simplest solution to housing pathogen-free rodents, but they require rigid discipline and operating procedures if they are to be used as a primary barrier against disease (Sedlacek et al., 1980).

HEPA-Filtered Laminar-Air-Flow Systems

These units are composed of modular chambers, hoods, and racks that place cages under a positive flow of filtered air, independent of the room ventilation system. They can either be used to house rodents or to hold rodents while cages are changed. Transportable units are commercially available (Figure 4-1); filters for whole-room application usually must be custom designed. Typically, these devices use high-efficiency particulate air (HEPA) filters that are capable of removing 99.97 percent of air-borne contaminants 0.3 μm or larger. A plenum and blower fan in the unit distribute the ultra-filtered air evenly over the cages in a horizontal laminar fashion. When planning an animal room with laminar-air-flow devices, careful consideration must be given to the energy requirements for multiple units and the heat load that is added to the room by the operation of these units. Low-velocity vertical mass air flow may be a successful adjunct to horizontal laminar air flow; however, the effectiveness of vertical mass air flow by itself is controversial (Thigpen and Ross, 1983). Regardless of the manufacturer or type of HEPA-filtered air supply device used in a room housing immunodeficient rodents, there are three essential components of management:

1. careful inspection and scheduled changing of the prefilters;
2. routine monitoring of the air velocity gauge, which indicates airflow rate through the HEPA filter (falling pressure requires corrective action); and
3. annual recertification of the HEPA filter unit by qualified service personnel to ensure that the HEPA filter is intact, properly seated, and not leaking.

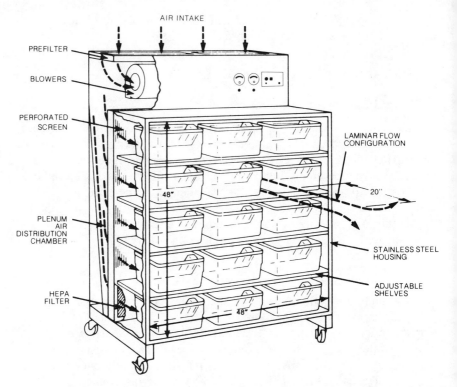

FIGURE 4-1 Diagram of a horizontal single-row laminar-air-flow cage rack for housing immunodeficient rodents. Figure courtesy of Lab Products, Inc., Maywood, N.J.

Attempts should be made to minimize dust in rooms equipped with these units to reduce the potential for loaded filters and failing systems.

Laminar-air-flow benches or cabinets, preferably on casters for easy management, can be used as a work surface for cage changing and experimental manipulation of animals. The principle applied here is the same as that for housing units. The committee considers this technique to be a useful adjunct to proper husbandry procedures, especially in facilities where barriers are not available or animal use is less restricted.

Germfree Isolators

For complete exclusion of microbes, the ideal housing system is the positive pressure isolator, such as those used to house germfree, or gnotobiotic, rodents. The most widely used isolators are made of flexible laminated vinyl plastic (ILAR, 1970, pp. 5–6). They can be chemically sterilized and easily adapted to specific needs. Isolators are also available in stainless steel, nylon, and polycarbonate (Plexiglas®). All such units have a filtered air supply and exhaust. Food and supplies must be sterilized and passed into the unit through a sterile entry chamber. Although isolators are an ideal housing system for keeping immunodeficient rodents pathogen-free, they are labor intensive, limit the number of animals that can be cared for in a given space, and seriously impair the ability to manipulate the animals.

Individually Ventilated Cage Racks

Ventilated racks that accommodate cages in enclosed cabinets or by suspension under closed tops and that supply air under positive or negative pressure or both are available. Depending on the intended use, these cage racks can supply HEPA-filtered air or animal room air. The efficacy of these systems in protecting immunodeficient animals from infectious agents has not been evaluated.

SPECIALIZED HUSBANDRY

General Considerations

The predominant consideration in the husbandry of immunodeficient rodents requiring protection is the very low resistance of these animals to infections of all types. The degree of protection provided should be dictated by the goals of the intended research. Precautions essential to gerontological studies, for example, might not be necessary in studies lasting only 1 or 2 months. However, all infections, including inapparent ones, are a constant threat to successful experimentation. Inapparent infections can, in a sense, be even more harmful than overt infections because they might go undetected and their impact on the study might be overlooked. Care must be taken to ensure that the results of an experiment in which immunodeficient rodents are used reflect the parameters measured and not concomitant infection.

Environmental Conditions

General rules that should be followed to ensure efficient husbandry are presented below.

Temperature

Immunodeficient rodents have been satisfactorily maintained in rooms where the temperature is regulated between 23.3°C (74°F) and 27.8°C (82°F). It has been shown that the thermoneutral zone for nude mice is above that for haired mice (Weihe, 1984) and that thyroid function in nude mice is impaired from birth (Pierpaoli and Besedovsky, 1975), resulting in hypothyroidism and poorly developed brown adipose tissue (Gripois et al., 1980). Therefore, it has been suggested that maintaining nude mice at 30–32°C (86.0–89.6°F) and providing thyroid hormone substitution could improve their health and life expectancy (Weihe, 1984). As a practical matter, these steps are rarely taken. Furthermore, excessively high room temperatures create husbandry problems, such as fermentation of feed and bedding, excess bacterial growth in watering systems, and an unpleasant work environment. Most managers find that a typical rodent room temperature of 21.1–23.3°C (70–74°F) plus filter caps, which the committee recommends universally for immunodeficient rodents, results in temperatures that are high enough for successful husbandry. Room temperature should be graphically monitored, and an alarm should be installed to signal a malfunction of the temperature control.

Humidity, Ventilation, and Lighting

Each of these factors has been specifically addressed in the *Guide for the Care and Use of Laboratory Animals* (NRC, 1985). Readers should refer to that document for details.

Food and Water

Food. Only sterilized or pasteurized diets should be fed because many commercial diets are contaminated with Enterobacteriaceae (Williams and Habermann, 1962). Sterilization of food reduces this potential source of infection. Vitamin-fortified, sterilizable diets are available and should be processed just prior to feeding. When diets are sterilized, the autoclave function must be carefully monitored by qualified personnel to ensure that sterilization has been achieved. It is also important that the vitamin content of the feed remains at recommended levels and that the diet is not excessively hard or clumped. Diets decontaminated by irradiation are also commercially available and can be considered as alternatives to steam-sterilized or pasteurized diets.

Water. Acidification of the drinking water with HCl to a pH of 2.5–2.8 has been found to be effective in controlling microbial contamination, including that with *Pseudomonas* spp. (McPherson, 1963). However, this procedure may impair the action of antibiotics and vitamins placed in the drinking water. In addition, excess acidification and chlorination have been associated with abnormalities in macrophage and lymphocyte function (Fidler, 1977; Hermann et al., 1982). The acidification procedure is still commonly used, however, because it is a practical way to control microbial contamination (Reed and Jutila, 1972).

Water bottles, stoppers, and drinking tubes should be washed and sterilized between uses.

Bedding

Bedding can also be a source of microbial contamination, and therefore, steam sterilization is recommended before use. Bedding material should not irritate the skin of rodents, especially those without hair. Ground corncob or fine hardwood chips are both acceptable, although the relatively poor absorbency of ground corncob bedding makes it less desirable. The addition of sterile, soft-tissue sheets to the cages of pregnant females for nesting material should be considered, especially for those strains that have a history of poor reproductive performance.

CONTROL OF INFECTION

Arrival of Stock from Commercial Sources and Transfer of Animals

Movement of animals into a facility for immunodeficient rodents should be held to an absolute minimum unless the facility is dependent on outside supply sources. The single most likely cause of barrier failure is the uncontrolled or ill-advised introduction of new animals into the barrier. In the case of barrier rooms operated as maintenance facilities for rodents from outside suppliers, the opportunities for failure exceed those of carefully managed, closed-colony facilities with integral breeding. In this case, it is essential that suppliers provide complete documentation of the health status of their colonies and that they be required to notify users immediately if they experience a failure of the desired health status of their colonies. When it is necessary to introduce new animals to provide genetic stock for backcrossing or other procedures, prior planning is mandatory. Germfree, specific-pathogen-free, defined-flora, or viral-antibody-free animals should be obtained if possible. On arrival, the containers must be examined for breaks and tears and the invoice must be examined to ensure that the animals are of the proper genetic and environmental history. Animals must be quarantined in a room separated from all other rodents until they are known to be free of disease.

Rodents being transferred into a facility must be tested for pathogenic bacteria, intestinal protozoa, and murine viruses (Hsu et al., 1980; Fox et al., 1984; Small, 1984). Some pathogenic organisms, such as *Pneumocystis carinii*, can only be detected by histological examination of cohorts or offspring of the stock to be introduced. In the case of athymic rodents, heterozygous littermates are ideal subjects for viral serology testing. If these are not available, gnotobiotic rodents or rodents from the barrier may be added to cages of prospective introductees to act as direct contact sentinels. These sentinels are examined serologically, 4–6 weeks later, for viral antibody titer. This procedure is sometimes referred to as cohabitation and is optimum when females are used as cohabitants to prevent fighting.

Regardless of the methods used to establish the health status of the new stock, they should be as thorough as possible and conducted in a quarantine holding area operated as a carefully managed barrier. Rodents transferred out of the colony should not be returned. Within the barrier, animal injection, anesthesia, and surgical procedures should all be conducted under aseptic conditions.

Disease Surveillance

Immunodeficient rodents must be observed constantly so that sick animals are detected promptly. Sick animals should be removed immediately to a quarantine area and submitted to the laboratory for diagnostic studies. Diagnostic tests to determine the cause of the illness should be conducted as soon as possible, and steps should be taken to prevent infection of the other animals. These tests should include complete necropsies, as well as bacteriologic and viral antibody assays. Because immunodeficient rodents are not able to react immunologically against all viruses (Eaton et al., 1975), they occasionally present a false-negative report. Therefore, an accurate evaluation requires that both immunodeficient and immunocompetent rodents be tested for the viral status of the colony. Immunocompetent sentinel rodents should be distributed randomly in a room that houses immunodeficient rodents, and a single individual from each cage of sentinels should be examined periodically for infectious diseases.

A complete discussion of diseases that commonly infect rodents can be found in ILAR (1974), Wagner and Manning (1976), Baker et al. (1979a), Foster et al. (1982), and Fox et al. (1984). Many of the infectious agents discussed in these documents produce far more severe manifestations of disease in immunodeficient rodents than they do in their normal counterparts.

5

Mating Systems for Mutants

There are many systems for breeding rodents, but not all of them can be used for the maintenance and propagation of mutants. The two major systems to be considered are *inbreeding*, which can be done either by brother × sister matings or by transferring a mutation to an inbred background, and *propagation without inbreeding*.

Some mutants can be inbred easily. Generally, this is the case when the mutation has no immediate deleterious effects and the affected animals of both sexes are fertile. A more involved process is necessary when the mutation reduces viability or fertility. Those systems of breeding used for most of the immunologic mutants discussed in this report are presented, starting with the simplest maintenance scheme and progressing to the most complex. Mouse mutants are used to illustrate each method; however, the mating systems are applicable to all rodents.

INBREEDING

Brother × Sister Inbreeding

Inbreeding of brother × sister littermates is the easiest system to use. At least 20 or more generations of consecutive brother × sister matings constitute an inbred strain. However, when expanding a strain beyond the F20 generation, all future matings must come from a common ancestor of at least the F20 generation, and, as inbreeding continues, the common ancestor should be kept as close as possible to the most advanced generation. The greatest hazard of inbreeding is the inbreeding depression (decreased fertility) that can occur.

Brother × Sister Inbreeding with Homozygosity

This type of inbreeding is done with recessive mutations when both male and female homozygotes are viable and fertile. This method is most useful for animals with mutations that cannot be visually distinguished at either pre- or postweaning ages. For example, lymphoproliferation (*lpr*) and generalized lymphoproliferative disease (*gld*) are not apparent until 8 and 12 weeks, respectively. Other immunologic mutations that cannot be visually distinguished include disease resistance and tolerance genes (*Bcg, Ity, Lsh,* and *Tol-1*), mitogen responsiveness (Lps^d), and immunodeficiency and complement genes (*scid, xid,* and *Hc*). These are all maintained in inbred strains that are homozygous for the specific immunologic gene.

Brother × Sister Inbreeding with Forced Heterozygosity

Heterozygosity (the condition of having one or more pairs of dissimilar alleles) can be forced upon a locus either by backcrossing or by intercrossing. This method can be used to produce an inbred background that is selected for expression of the specific mutation or to maintain a mutation on an already inbred background when it is desirable to have a nonmutant sibling as a control. Backcrosses can be used if the gene is recessive and viable (i.e., r/+ × r/r) or if the gene is dominant (i.e., D/+ × +/+). Intercrosses can be used if the gene is recessive and lethal or sterile (i.e., r/+ × r/+) or if the gene is dominant or semidominant and lethal (i.e., D/+ × D/+) (E. L. Green, 1966). The mutations discussed in this report that are maintained in this manner are hairless (*hr*), beige (*bg*), and dwarf (*dw*). Strain HRS/J is an example of an inbred strain produced by brother × sister matings of a haired (*hr/+*) female with a hairless (*hr/hr*) male. Strain DW/J is an example of an inbred strain produced by brother × sister matings of heterozygotes (*dw/+*). The untreated homozygote (*dw/dw*) is sterile. However, because heterozygotes are phenotypically indistinguishable from homozygous normal animals, each pair of normal-appearing siblings must produce *dw/dw* offspring before that pair is placed in the breeding colony.

Inbreeding a Balanced Stock

Closely linked mutant genes can be used to distinguish heterozygotes of lethal or sterile recessive mutants. Because these lethal or sterile mutants must be bred from heterozygotes, there is a considerable saving of cage space if the heterozygote can be recognized at least 70 percent of the time. To make a balanced stock, a carrier of the recessive mutation in question ($m/+$) is crossed to a mouse that is homozygous or heterozygous for a linked marker gene (g/g or $g/+$). The offspring are then intercrossed, and, if both mutant phenotypes (m/m and g/g) are among their progeny, the intercrossed parents are double heterozygotes ($m\ +/+\ g$) in the repulsion phase; that is, the two recessive mutant alleles are on different members of the homologous chromosome pair.

From double heterozygote repulsion matings, three classes of offspring can be distinguished: homozygotes of the gene in question; homozygotes of the closely linked marker gene; and normal-appearing mice, which are expected at least 70 percent of the time to be carriers of both genes if the genes are linked as close as 7 cM (M. C. Green, 1966). Only normal-appearing offspring from parents that have produced both mutants should be used to propagate the next generation (for more details regarding this type of mating see M. C. Green, 1966). Examples of stocks that are maintained in this manner are grey-lethal (gl) and diabetes (db). The grey-lethal mutation in strain GL/Le is balanced with downless-J (dl^J), which is located approximately 7 cM away from gl on chromosome 10. Diabetes (db) on C57BL/Ks is maintained balanced with misty (m), which is about 1 cM away on chromosome 4.

Closely linked marker genes can also be used in coupling (both mutant alleles are on the same chromosome) to help identify mutants before the effect of the gene can be detected.

Transferring a Mutation to an Inbred Background

One of the standard inbred strains (e.g., C57BL/6J or C3H/HeJ) is used to provide the background for this system of breeding. The mutant gene is crossed into the selected strain in one of the following ways, depending on whether the mutation is dominant, recessive, or recessive and lethal or sterile.

Backcross Matings

Backcross matings are generally used for a dominant gene such as viable dominant spotting (W^v). The heterozygote ($W^v/+$) is repeatedly backcrossed to a member of the selected inbred strain ($+/+$). This backcrossing continues for several generations. The letter "N" is used to denote the number of times a mutation has been crossed to the inbred background. Between N7 and N10, nearly all alleles that are not closely linked to the W locus will have come from the selected inbred strain (E. L. Green, 1966).

Cross–Intercross Matings Using Homozygotes

Cross–intercross matings, using animals homozygous for the recessive gene, are generally used when the homozygote is viable and fertile. In this type of mating, the homozygous recessive mutant (r/r) is crossed to the selected inbred strain ($+/+$). The offspring are mated brother × sister (intercrossed = $r/+$ × $r/+$), and the homozygote (r/r) is recovered and crossed back again to the inbred strain ($+/+$). This pattern of cross–intercross matings is continued until at least 7–10 crosses (N7 to N10) back to the inbred strain have been completed. At this point, crossing into the inbred strain can continue or brother × sister matings can be made using either backcross- or intercross-type matings. However, if the brother × sister method is used, it is wise to occasionally cross back to the inbred strain to prevent subline divergence. The mutation *lpr* has been put on several different inbred backgrounds to at least the N10 generation by this method and then maintained by brother × sister matings of homozygotes.

Cross–Intercross Matings Using Heterozygotes

Cross–intercross matings using heterozygotes are made when the recessive gene is lethal or sterile. A known heterozygote ($r/+$) is crossed to the selected inbred strain ($+/+$). Only one out of two of the offspring, or one out of four pairs, is expected to be a carrier; therefore, as many as 12 brother × sister pairs might have to be made up from the progeny to ensure that two carriers are mated. Carriers are identified by the production of mutant offspring. Once identified, the heterozygote ($r/+$) is crossed to the inbred strain ($+/+$), and the progeny are again intercrossed. This breeding pattern can be continued indefinitely, or after the mutation has been placed on the inbred background with at least seven crosses (N7) to ensure histocompatibility, the ovarian transplantation technique can be used. Ovarian transplantation is a more efficient method of maintenance because all intercross pairs are known carriers, and, consequently, fewer pairs are needed to maintain the stock.

Cross–Intercross Matings Using Ovarian Transplantation

Cross–intercross matings using ovarian transplantation are used to maintain lethal or sterile recessive mutations. The ovaries of the homozygous mutant are removed at any time from 14 days to several weeks of age and are transplanted into the empty ovarian capsule of a histocompatible female host. The host must be of such a genotype (generally a coat color is used) that the appearance of the offspring will determine whether they were produced from eggs from the transplanted ovary or from residual tissue in the host ovary. The host female is then crossed to a male of the selected inbred strain. For example, if a lethal mutant gene arose on the C57BL/6J strain, it will be nonagouti (a/a) in color. The host might be a white-bellied agouti female from the coisogenic C57BL/6J-A^{w-J}/A^{w-J} strain or a hybrid of C57BL/6J-A^{w-J}/A^{w-J} × any inbred strain that is homozygous agouti (A/A) in color. After ovarian transplantation, this host is crossed to a C57BL/6J male that is nonagouti (a/a). If the offspring are of the correct color (a/a), they will be intercrossed (brother × sister mated) and a homozygous mutant offspring will be selected as the ovarian donor for the next transplant. If the offspring are agouti in color, they are from the host ovary and are discarded. The new host with the nonagouti mutant ovary is then crossed to the inbred C57BL/6J strain. Each cross to the inbred strain represents another N generation. However, before ovarian transplantation to an inbred strain can be made, the lethal or sterile recessive mutation must be placed on the selected inbred strain to N7 by the cross–intercross system using the heterozygote to ensure histocompatibility (unless it arose on that strain by mutation and is, therefore, coisogenic and histocompatible). The mutations motheaten (me), viable motheaten (me^v), microphthalmia (mi), and obese (ob) are all maintained on inbred strains by ovarian transplantation.

PROPAGATION WITHOUT INBREEDING

Some mutant mice cannot be successfully inbred; consequently, the use of the hybrid mouse has become an invaluable means for both maintaining and producing these mutants. In general, the vigor of the hybrid results in hardier, faster-growing, and longer-lived mutants, as well as better reproductive performance in the breeders.

Transferring a Mutation to a Hybrid Background

A mutation can be transferred to a hybrid background in two ways. First, the mutation can be transferred onto two different standard inbred strains. Hybrid mutant mice can then be produced by crossing one mutant-bearing strain to the other. The mutant from this cross is a true hybrid or F1. However, this is an expensive and space-consuming process, and, although hybrid mutants and controls with known genotypes are produced, the breeding stock is inbred and is often very difficult to maintain.

The second method is to transfer the mutation to a hybrid background that is made from two standard inbred strains and that is also color coded to permit ovarian transplantation. For example, strain C57BL/6J is nonagouti (a/a), and the coisogenic strain C57BL/6J-A^{w-J}/A^{w-J} is white-bellied agouti. Strain C3HeB/FeJ is agouti (A/A), and the congenic strain C3HeB/FeJLe-a/a is nonagouti. From these four strains two sets of compatible hybrids can be made, a nonagouti, B6C3Fe-a/a F$_1$ (C57BL/6J × C3HeB/FeJLe-a/a F$_1$), and an agouti, C3FeB6-A/A^{w-J} F$_1$ (C3HeB/FeJ × C57BL/6J-A^{w-J}/A^{w-J}F$_1$).

The type of cross used to transfer a mutation onto a hybrid background depends on whether the mutation is dominant, recessive, or recessive and lethal or sterile.

Backcross Matings

Backcross matings to the hybrid can be made with a dominant gene just as they are made to an inbred strain. Dominant hemimelia (Dh) is maintained in this manner.

Cross–Intercross Matings

Cross–intercross matings can be made with recessive genes that are either homozygous or heterozygous to any hybrid in the same manner that a recessive gene is transferred to a standard inbred strain. Lethargic (lh) is maintained on the B6C3-a/A F_1 (C57BL/6J × C3H/HeSnJ F_1) background in this manner.

Cross–Intercross Matings Using Ovarian Transplantation

Cross–intercross matings using ovarian transplantation can be accomplished using the compatible sets of hybrids just described. For example, the mutation osteopetrosis (op) is recessive, and the homozygote does not breed. It is maintained by crossing to the nonagouti, B6C3Fe-a/a F_1 hybrid via ovarian transplantation and intercrossing those progeny. However, before ovarian transplantation can be done, the mutant gene must first be transferred to this hybrid background by at least seven crosses (N7) using a heterozygote ($op/+$). Once the op/op mouse is compatible with the B6C3Fe-a/a F_1 background, an a/a op/op ovary is transplanted into the ovarian capsule of the C3FeB6-A/A^{w-J} F_1 female. This host is crossed to a B6C3Fe-a/a F_1 male, and all nonagouti (a/a) offspring will be $op/+$ and can be intercrossed. Any agouti offspring are from host tissue and can be discarded. This intercross is expected to produce one out of four affected (op/op) offspring, out of which a female that is a/a op/op is again selected for an ovarian transplant. Each cross to the B6C3Fe-a/a F_1 male is another N generation. Osteopetrosis (op), osteosclerosis (oc), and wasted (wst) are maintained in this manner.

Because of the hybrid vigor, the mutants are generally husky and the breeders have large litters; therefore, many heterozygotes can be produced in a very limited space. This is the most efficient method of maintaining lethal and sterile mutations that cannot be successfully inbred.

6

Genetic Mechanisms Governing Resistance or Susceptibility to Infectious Diseases

Specific immunity has been associated with protection against most forms of infectious agents in rodents and other vertebrate animals. The role of MHC genes in this specific immunity has received much attention. It is the subject of many review articles (Benacerraf and Rock, 1984; David, 1984; Zinkernagel et al., 1985; Ashman, 1987; Van Bleek et al., 1988) and will not be discussed in this section. Likewise, a review of the molecular genetics of retroviruses (oncornaviruses), which has been addressed elsewhere in great detail, is beyond the scope of this report. Those interested in learning more about the genetic regulation and expression of retroviruses are urged to read any of a number of current reviews on the subject (Pincus, 1980; Kulstad, 1986; Salzman, 1986). This chapter deals with susceptibility and resistance to infectious agents not associated with specific immunity or linked to the MHC.

The resistance of a host to an infectious agent is controlled by several nonspecific factors, including age, gender, diet, heredity, hormone status, metabolic changes, macrophage function, and the inhibitory substances interferon and thymosin. In addition, resistance or susceptibility is relative to and dependent on strain, dose, and route of inoculation of the infectious agent. Various strains of mice are known to differ in their susceptibilities to naturally occurring viral agents such as Sendai, ectromelia, polyoma, and lactic dehydrogenase viruses (Bang and Warwick, 1960; Chang and Hildemann, 1964; Kees and Blanden, 1976; Martinez et al., 1980). Strain differences are also observed in their susceptibilities to such experimental virus infections as herpes simplex type 1, influenza A, measles, and flaviviruses

(Sabin, 1952; Lindemann, 1964; Lopez, 1975; Rager-Zisman et al., 1980). In each of these instances, genes outside the MHC appear to be important determinants of susceptibility. A single autosomal dominant trait not linked to *H-2* (actually it is found on chromosome 1) confers resistance to Sendai virus infection in C57BL/6J mice (Brownstein, 1983). Both *H-2* and non-*H-2* loci have been found to confer resistance to the mouse coronavirus MHV$_3$, and it has been postulated that the genes conferring this resistance operate through macrophages and cells that mount an immune response (Dupuy et al., 1984). The resistance of C57BL/6 mice to the demyelinating disease caused by the JHM strain of MHV has been shown to be dependent on T-cell activity and not interferon (Sorensen et al., 1982); however, it is also greatly influenced by the age of the host (Pickel et al., 1981).

In rats, a demyelinating disease associated with autoimmunity can be induced by the JHM strain of MHV. Intracerebral infection of LEW rats can lead to encephalomyelitis (Nagashima et al., 1979). BN rats do not develop the disease; a strong, local, virus-specific antibody response effectively controls the intracerebral spread of JHM virus (Dörries et al., 1987). The demyelinating disease is largely caused by a T-cell-mediated delayed-type hypersensitivity reaction against brain tissue. Adoptive transfer from diseased LEW rats of lymphocytes restimulated in vitro with basic myelin protein leads to lesions resembling an experimental allergic encephalomyelitis in recipients (Watanabe et al., 1983). Host factors such as strain and age affect the development of the disease (Wege et al., 1987).

A gene outside the MHC (*RT1*) appears to control suceptibility or resistance. Preliminary data map this gene into linkage group V of the rat (H. J. Hedrich, Central Institute for Laboratory Animal Breeding, Hannover, Federal Republic of Germany, unpublished data). Ectromelia virus (mousepox) has been extensively studied in various inbred strains of mice. When inoculated into the footpad, marked differences in the 50 percent lethal dose (LD$_{50}$) have been observed among strains, with C57BL mice being the most resistant (O'Neill and Blanden, 1983). Both *H-2* and non-*H-2* genes appear to be responsible for this resistance, and studies in chimeric mice have demonstrated that the superior resistance of C57BL/6 over BALB/c mice operates through reduced early (1–2 days) virus transmission to the lymphoreticular system by a radioresistant cell or substance. Noteworthy is the report of Wallace and Buller (1985) that a strain of ectromelia virus isolated in 1979 from an outbreak at the National Institutes of Health could be passaged up to seven times by exposure to infected C57BL/6J cage mates without any apparent clinical illness. Furthermore, these authors found that females are significantly more resistant than males. These studies are in marked contrast to earlier ones in which the Moscow strain of ectromelia virus was used. Outbred stocks of mice were exquisitely sensitive to this strain of the virus, and the virus was invariably fatal (Fenner, 1949).

The X-linked resistance of mice to high doses of herpes simplex virus type 2 has been found to correlate with interferon production (Pedersen et al., 1983). By contrast, the increased susceptibility of male mice to inoculation with herpes simplex virus type 1 is thought to involve antibody-depressing mechanisms of androgen-sensitive cell populations (Knoblich et al., 1983). BALB/c mice carry genes that impart protection against lethal encephalomyocarditis virus infection by influencing the action of interferon (Dandoy et al., 1982). In C3H/RV mice, however, an autosomal dominant gene confers resistance to flavivirus infection that is independent of interferon action (Brinton et al., 1982). Differential susceptibility to the demyelinating disease associated with Theiler's murine encephalomyelitis virus is controlled by at least two non-H-2 genes in mice (Melvold et al., 1987). Strain differences in susceptibility to viral agents have also been described in rats (Sorensen et al., 1982), hamsters (Fultz et al., 1981), and guinea pigs (Jahrling et al., 1982).

The severity of pulmonary lesions in mice and rats inoculated intranasally with *Mycoplasma pulmonis* has been studied (Cassell, 1982). Saito et al. (1978b) found that ddY (not standardized nomenclature) mice are less resistant to respiratory disease than are ICR mice. Likewise, LEW rats are more susceptible to severe pulmonary disease caused by *M. pulmonis* than are F344 rats when age, sex, and environmental factors are controlled (Davis et al., 1982). In addition, the infection persists in LEW rats for 120 days, as opposed to 28 days in F344 rats. It has been suggested that *M. pulmonis* is a B-cell activator in LEW rats, and differences in the severity of infection can be attributed to differences in nonspecific lymphocyte activation (Naot et al., 1979).

Resistance and susceptibility to other infectious and parasitic agents have also been examined. Jerrells and Osterman (1982) found that resistance to intraperitoneal infection with *Rickettsia tsutsugamushi* in C3H/RV mice was not radiation sensitive and most likely concerned differences in Ia-positive macrophages (Jerrells, 1983). Similarly, the resistance of mice to infection with *Rickettsia akari* was found to reside in macrophages and was determined by loci other than Lps^d (Nacy and Meltzer, 1982).

Morozumi and others (1981) found that the susceptibility of inbred strains of mice to *Blastomyces dermatitidis* was not related to cellular or humoral immunity but rather to macrophage activity controlled by a non-Lps locus. Likewise, Kirkland and Fierer (1983) found that murine resistance to *Coccidioides immitis* is controlled as a dominant trait and is not associated with the *Lsh* or *Lps* loci or with the genes controlling resistance to *Blastomyces dermatitidis*. Murine resistance to paracoccidioidomycosis was found to be controlled by the autosomal dominant gene *Pbr*, which is not linked to *Cms*, the resistance gene for coccidioidomycosis (Calich et al., 1987).

R. G. Bell and colleagues (1984) discovered that the rapid expulsion of

Trichinella spiralis in mice is controlled by an autosomal dominant gene called *Ihe-1* (intestinal helminth expulsion-1), which is not linked to the MHC, *Lsh*, or chromosome 7. This gene, *Ihe-1*, was not linked to genes controlling resistance to *Taenia teaniaformis*, *Giardia muris*, *Trichuris muris*, or *Nematospiroides dubius*.

Although the mechanisms controlling resistance and susceptibility to the agents listed above are, for the most part, undefined, it is known that they do not involve specific immunity. As additional loci are identified that control various aspects of the immune response, some of these resistance genes might be found to have a basis in specific immunity.

References

Abbott, C. M., C. J. Skidmore, A. G. Searle, and J. Peters. 1986. Deficiency of adenosine deaminase in the wasted mouse. Proc. Natl. Acad. Sci. USA 83:693–695.
Accinni, L., and F. J. Dixon. 1979. Degenerative vascular disease and myocardial infarction in mice with lupus-like syndrome. Am. J. Pathol. 96:477–492.
Acha-Orbea, H., and H. O. McDevitt. 1987. The first external domain of the nonobese diabetic mouse class II I-A β chain is unique. Proc. Natl. Acad. Sci. USA 84:2435–2439.
Akizuki, M., J. P. Reeves, and A. D. Steinberg. 1978. Expression of autoimmunity by NZB/NZW marrow. Clin. Immunol. Immunopathol. 10:247–250.
Alberts, J. R. 1974. Producing and interpreting experimental olfactory deficits. Physiol. Behav. 12:657–670.
Alcock, N. W., and M. E. Shils. 1974. Serum immunoglobulin G in the magnesium-depleted rat. Proc. Soc. Exp. Biol. Med. 145:855–858.
Alexander, E. L., E. D. Murphy, J. B. Roths, and G. E. Alexander. 1983. Congenic autoimmune murine models of central nervous system disease in connective tissue disorders. Ann. Neurol. 14:242–248.
Allison, A. C. 1974. Interactions of antibodies, complement components and various cell types in immunity against viruses and pyogenic bacteria. Transplant. Rev. 19:3–55.
Alper, C. A., and F. S. Rosen. 1984. Inherited deficiencies of complement proteins in man. Springer Semin. Immunopathol. 7:251–261.
Altman, A., A. N. Theofilopoulos, R. Weiner, D. H. Katz, and F. J. Dixon. 1981. Analysis of T cell function in autoimmune murine strains. Defects in production of and responsiveness to interleukin 2. J. Exp. Med. 154:791–808.
Amagai, T., T. Matsuzawa, and B. Cinader. 1982. The effect of ribavirin and rifamycin SV on age-dependent changes of the immune system during the adult life of SJL mice. Immunol. Lett. 4:149–153.
American Veterinary Medical Association (AVMA) Panel on Euthanasia. 1986. 1986 Report of the AVMA Panel on Euthanasia. J. Am. Vet. Med. Assoc. 188:252–268.
Amornsiripanitch, S., L. M. Barnes, J. J. Nordlund, L. S. Trinkle, and L. A. Rheins. 1988.

Immune studies in the depigmenting C57BL/Ler-*vit/vit* mice. An apparent isolated loss of contact hypersensitivity. J. Immunol. 140:3438–3445.

Amsbaugh, D. F., C. T. Hansen, B. Prescott, P. W. Stashak, D. R. Barthold, and P. J. Baker. 1972. Genetic control of the antibody response to type III pneumococcal polysaccharide in mice. I. Evidence that an X-linked gene plays a decisive role in determining responsiveness. J. Exp. Med. 136:931–949.

Andrews, B. S., R. A. Eisenberg, A. N. Theofilopoulos, S. Izui, C. B. Wilson, P. J. McConahey, E. D. Murphy, J. B. Roths, and F. J. Dixon. 1978. Spontaneous murine lupus-like syndromes. Clinical and immunopathological manifestations in several strains. J. Exp. Med. 148:1198–1215.

Andriole, G. L., J. J. Mulé, C. T. Hansen, W. M. Linehan, and S. A. Rosenberg. 1985. Evidence that lymphokine-activated killer cells and natural killer cells are distinct based on an analysis of congenitally immunodeficient mice. J. Immunol. 135:2911–2913.

Apte, R. N., O. Ascher, and D. H. Pluznik. 1977. Genetic analysis of generation of serum interferon by bacterial lipopolysaccharide. J. Immunol. 119:1898–1902.

Archer, D. L., B. G. Smith, and H. M. Johnson. 1981. Direct and induced suppression of in vitro antibody production by selected phenols: A comparison to virus-type interferon. Pp. 199–209 in Biological Relevance of Immune Suppression as Induced by Genetic, Therapeutic, and Environmental Factors, J. H. Dean and M. Padarathsingh, eds. Proceedings of a technology assessment workshop held November 13–14, 1979, in Williamsburg, Va. New York: Van Nostrand Reinhold.

Archinal, W. A., and M. S. Wilder. 1988a. Susceptibility of HRS/J to listeriosis: Dynamics of infection. Infect. Immun. 56:607–612.

Archinal, W. A., and M. S. Wilder. 1988b. Susceptibility of HRS/J to listeriosis: Macrophage activity. Infect. Immun. 56:613–618.

Arroyave, C. M., R. M. Levy, and J. S. Johnson. 1974. Genetic deficiency of the fourth component of complement in Wistar rats. Fed. Proc. 33:795 (abstr.).

Arroyave, C. M., R. M. Levy, and J. S. Johnson. 1977. Genetic deficiency of the fourth component of complement (C4) in Wistar rats. Immunology 33:453.

Ash, P., J. F. Loutit, and K. M. S. Townsend. 1980. Osteoclasts derived from haematopoietic stem cells. Nature (London) 283:669–670.

Ashman, R. B. 1987. Murine candidiasis. III. Host inflammatory responses are regulated in part by class I MHC genes. J. Immunogenet. 14:317–321.

Atkinson, J. P., K. McGinnis, L. Brown, J. Peterein, and D. Shreffler. 1980. A murine C4 molecule with reduced hemolytic efficiency. J. Exp. Med. 151:492–497.

Aubert, R., J. Herzog, M.-C. Camus, J.-L. Guenet, and D. Lemonnier. 1985. Description of a new model of genetic obesity: db^{Pas}. J. Nutr. 115:327–333.

Auerbach, H. S., R. Burger, D. Bitter-Suermann, G. Goldberger, and H. R. Colten. 1985. C3-deficient guinea pig mRNA directs synthesis of a structurally abnormal C3 protein. Complement 2:5 (abstr.).

Averdunk, R., P.-H. Bippus, T. Günther, and H. J. Merker. 1982. Development and properties of malignant lymphoma induced by magnesium deficiency in rats. J. Cancer Res. Clin. Oncol. 104:63–73.

Azar, H. A., C. T. Hansen, and J. Costa. 1980. N:NIH(S)II-*nu/nu* mice with combined immunodeficiency: A new model for human tumor heterotransplantation. J. Natl. Cancer Inst. 65:421–430.

Bach, J.-F., M. Dardenne, and J.-C. Salomon. 1973. Studies on thymus products. IV. Absence of serum 'thymic activity' in adult NZB and (NZB × NZW) F1 mice. Clin. Exp. Immunol. 14:247–256.

Bach, M.-A., D. Droz, L.-H. Noel, D. Blanchard, M. Dardenne, and A. Peking. 1980. Effect

of long-term treatment with circulating thymic factor on murine lupus. Arthritis Rheum. 23:1351–1358.
Baekkeskov, S., T. Dyrberg, and A. Lernmark. 1984. Autoantibodies to a 64-kilodalton islet cell protein precede the onset of spontaneous diabetes in the BB rat. Science 224:1348–1350.
Baetjer, A. M. 1968. Role of environmental temperature and humidity in susceptibility to disease. Arch. Environ. Health 16:565–570.
Bailey, D. W. 1979. Genealogies of long-separated sublines in six major inbred mouse strains. Pp. 18–20 in Inbred and Genetically Defined Strains of Laboratory Animals. Part 1. Mouse and Rat, P. L. Altman and D. D. Katz, eds. Bethesda, Md.: Federation of American Societies for Experimental Biology.
Baker, H. J., G. H. Cassell, and J. R. Lindsey. 1971. Research complications due to *Haemobartonella* and *Eperythrozoon* infections in experimental animals. Am. J. Pathol. 64:625–656.
Baker, H. J., J. R. Lindsey, and S. H. Weisbroth, eds. 1979a. The Laboratory Rat. Vol. I: Biology and Diseases. New York: Academic Press. 435 pp.
Baker, H. J., J. R. Lindsey, and S. H. Weisbroth. 1979b. Housing to control research variables. Pp. 169–192 in The Laboratory Rat. Vol. 1: Biology and Diseases, H. J. Baker, J. R. Lindsey, and S. H. Weisbroth, eds. New York: Academic Press.
Bancroft, G. J., M. J. Bosma, G. C. Bosma, and E. R. Unanue. 1986. Regulation of macrophage Ia expression in mice with severe combined immunodeficiency: Induction of Ia expression by a T cell-independent mechanism. J. Immunol. 137:4–9.
Bang, F. B., and A. Warwick. 1960. Mouse macrophages as host cells for the mouse hepatitis virus and the genetic basis of their susceptibility. Proc. Natl. Acad. Sci. USA 46:1065–1075.
Bannerman, R. M., J. A. Edwards, and P. H. Pinkerton. 1973. Hereditary disorders of the red cell in animals. Prog. Hematol. 8:131–179.
Barkley, M. S., A. Bartke, D. S. Gross, and Y. N. Sinha. 1982. Prolactin status of hereditary dwarf mice. Endocrinology 110:2088–2096.
Barnes, A. D., and M. Tuffrey. 1967. Serum antinuclear factor and the influence of environment in mice. Nature (London) 214:1136–1138.
Baroni, C. 1967. Thymus, peripheral lymphoid tissues and immunological responsiveness of the pituitary dwarf mouse. Experientia 23:282–283.
Baroni, C., N. Fabris, and G. Bertoli. 1967. Age dependency of the primary immune response in the hereditary pituitary dwarf and normal Snell-Bagg mouse. Experientia 23:1059–1060.
Baroni, C. D., N. Fabris, and G. Bertoli. 1969. Effects of hormones on development and function of lymphoid tissues. Synergistic action of thyroxin and somatotropic hormone in pituitary dwarf mice. Immunology 17:303–314.
Bartke, A. 1964. Histology of the anterior hypophysis, thyroid and gonads of two types of dwarf mice. Anat. Rec. 149:225–235.
Bean-Knudsen, D. E., and J. E. Wagner. 1987. Effect of shipping stress on clinicopathologic indicators in F344/N rats. Am. J. Vet. Res. 48:306–308.
Bearer, E., M. E. Gershwin, and J. J. Castles. 1986. Lupus—Insights from experimental models. Clin. Exp. Rheumatol. 4:161–168.
Bedigian, H. G., L. D. Shultz, and H. Meier. 1979. Expression of endogenous murine leukaemia viruses in AKR/J streaker mice. Nature (London) 279:434–436.
Behar, S. M., and M. D. Scharff. 1988. Somatic diversification of the S107 (T15) V_H11 germline gene that encodes the heavy-chain variable region of antibodies to double-stranded DNA in (NZB × NZW) F1 mice. Proc. Natl. Acad. Sci. USA 85:3970–3974.
Behlke, M. A., H. S. Chou, K. Huppi, and D. Y. Loh. 1986. Murine T-cell receptor mutants with deletions of β-chain variable region genes. Proc. Natl. Acad. Sci. USA 83:767–771.

Beisel, K. W., and S. P. Lerman. 1981. Progressive loss of H-2D^s antigens during passage of a spontaneous SJL/J lymphoma. Transplant. Proc. 13:1811–1813.

Bell, M., L. A. Rheins, and J. J. Nordlund. 1984. Cytologic alterations in the epidermis of the depigmenting C57BL-*vi/vi* mouse. J. Cell Biol. 99(4, part 2):386a (abstr. 1424).

Bell, R. G., L. S. Adams, and R. W. Ogden. 1984. A single gene determines rapid expulsion of *Trichinella spiralis* in mice. Infect. Immun. 45:273–275.

Bellgrau, D., A. Naji, W. K. Silvers, J. F. Markmann, and C. F. Barker. 1982. Spontaneous diabetes in BB rats: Evidence for a T-cell dependent immune response defect. Diabetologia 23:359–364.

Benacerraf, B., and K. L. Rock. 1984. Cellular expression and function of Ir genes. Pp. 13–20 in Immunogenetics. Its Application to Clinical Medicine, T. Sasazuki and T. Tada, eds. Tokyo: Academic Press.

Bendelac, A., C. Carnaud, C. Boitard, and J. F. Bach. 1987. Syngeneic transfer of autoimmune diabetes from diabetic NOD mice to healthy neonates. Requirement for both L3T4$^+$ and Lyt-2$^+$ T cells. J. Exp. Med. 166:823–832.

Benjamin, W. H., Jr., and D. E. Briles. 1982. The net growth of Aro A$^-$ mutant *Salmonella typhimurium* and the growth of wild type *S. typhimurium* in mice are under different genetic control. Abstr. B-107, p. 35 in Abstracts of the Annual Meeting of the American Society for Microbiology, 1982. (Available from the Meetings Department, American Society for Microbiology, 1913 I Street, NW, Washington, DC 20006.)

Benjamin, W. H., Jr., S. J. Roberts, and D. Briles. 1987. The use of a temperature sensitive plasmid to show that the mouse *Ity^r* gene causes decreased growth of *Salmonella typhimurium* in mice. Abstr. B-97, p. 41 in Abstracts of the Annual Meeting of the American Society for Microbiology, 1987. (Available from the Meetings Department, American Society for Microbiology, 1913 I Street, NW, Washington, DC 20006.)

Bennett, M., and R. A. Steeves. 1970. Immunocompetent cell functions in mice infected with Friend leukemia virus. J. Natl. Cancer Inst. 44:1107–1119.

Bentwich, Z., B. Bonavida, A. Peled, and N. Haran-Ghera. 1972. Evaluation of autoantibodies and serum abnormalities in SJL/J mice. Isr. J. Med. Sci. 8:670 (abstr.).

Ben-Yaakov, M., and N. Haran-Ghera. 1975. T and B lymphocytes in thymus of SJL/J mice. Nature (London) 255:64–66.

Bernard, C. C. A. 1976. Experimental autoimmune encephalomyelitis in mice: Genetic control of susceptibility. J. Immunogenet. 3:263–274.

Berridge, M. V., N. O'Kech, L. J. McNeilage, B. F. Heslop, and R. Moore. 1979. Rat mutant (*NZNU*) showing "nude" characteristics. Transplantation 27:410–413.

Besedovsky, H. O., A. Del Ray, E. Sorkin, W. Lotz, and U. Schwulera. 1985. Lymphoid cells produce an immunoregulatory glucocorticoid increasing factor (GIF) acting through the pituitary gland. Clin. Exp. Immunol. 59:622–628.

Bhatt, P. N., and R. O. Jacoby. 1986. Mousepox: Pathogenesis, diagnosis and rederivation. Pp. 557–570 in Viral and Mycoplasmal Infections of Laboratory Rodents. Effects on Biomedical Research, P. N. Bhatt, R. O. Jacoby, H. C. Morse III, and A. E. New, eds. Proceedings of a conference held October 24–26, 1984, in Bethesda, Md. Orlando, Fla.: Academic Press.

Bielschowsky, M., F. Bielschowsky, and D. Lindsay. 1956. A new strain of mice with a high incidence of mammary cancers and enlargement of the pituitary. Br. J. Cancer 10:688–699.

Bielschowsky, M., B. J. Helyer, and J. B. Howie. 1959. Spontaneous hemolytic anemia. Proc. Univ. Otago Med. School 37:9–11.

Biggar, R. J., R. Diebel, and J. P. Woodall. 1976. Implications, monitoring, and control of accidental transmission of lymphocytic choriomeningitis virus within hamster tumor cell lines. Cancer Res. 36:551–553.

Billings, P. B., R. W. Allen, F. C. Jensen, and S. O. Hoch. 1982. Anti-RNP monoclonal

antibodies derived from a mouse strain with lupus-like autoimmunity. J. Immunol. 128:1176–1180.
Bitter-Suermann, D., T. Hoffmann, R. Burger, and U. Hadding. 1981. Linkage of total deficiency of the second component (C2) of the complement system and of genetic C2-polymorphism to the major histocompatibility complex of the guinea pig. J. Immunol. 127:608–612.
Black, P. L., M. Holly, C. I. Thompson, and D. L. Margules. 1983. Enhanced tumor resistance and immunocompetence in obese (ob/ob) mice. Life Sci. 33(suppl. 1):715–718.
Blackwell, J. M. 1985. A murine model of genetically controlled host responses to leishmaniasis. Pp. 147–155 in Ecology and Genetics of Host-Parasite Interactions, D. Rollinson and R. M. Anderson, eds. Linnean Soc. Symp. Series, no. 11. Papers presented at an international symposium organized by the Linnean Society of London and the British Society for Parasitology, held July 12–13, 1984, at Keele University. London: Academic Press.
Blaineau, C., S. Gisslebrecht, D. Piatier, M.-A. Hurot, and J. P. Levy. 1978. Swan mice: Autoimmune mice without xenotropic type C virus production. Clin. Exp. Immunol. 31:276–280.
Blakley, B. R., C. S. Sisodia, and T. K. Mukkur. 1980. The effect of methylmercury, tetraethyl lead, and sodium arsenite on the humoral immune response in mice. Toxicol. Appl. Pharmacol. 52:245–254.
Bleby, J. 1976. Disease-free (SPF) animals. Pp. 123–134 in The UFAW Handbook on the Care and Management of Laboratory Animals, 5th ed., Universities Federation for Animal Welfare (UFAW), ed. Edinburgh: Churchill Livingstone.
Blomgren, H., and B. Andersson. 1976. Steroid sensitivity of the PHA and PWM responses of fractionated human lymphocytes in vitro. Exp. Cell Res. 97:233–240.
Boillot, D., R. Assan, M. Dardenne, M. Debray-Sachs, and J. F. Bach. 1986. T-lymphopenia and T-cell imbalance in diabetic db/db mice. Diabetes 35:198–203.
Bois, P., E. B. Sandborn, and P. E. Messier. 1969. A study of thymic lymphosarcoma developing in magnesium-deficient rats. Cancer Res. 29:763–775.
Boitard, C., S. Michie, P. Serrurier, G. W. Butcher, A. P. Larkins, and H. O. McDevitt. 1985. In vivo prevention of thyroid and pancreatic autoimmunity in the BB rat by antibody to class II major histocompatibility complex gene products. Proc. Natl. Acad. Sci. USA 82:6627–6631.
Bonnard, G. D., E. K. Manders, D. A. Campbell, Jr., R. B. Herberman, and M. J. Collins, Jr. 1976. Immunosuppressive activity of a subline of the mouse EL-4 lymphoma. Evidence for minute virus of mice causing the inhibition. J. Exp. Med. 143:187–205.
Boorman, G. A., M. I. Luster, J. H. Dean, M. L. Campbell, L. A. Lauer, F. A. Talley, R. E. Wilson, and M. J. Collins. 1982. Peritoneal macrophage alterations caused by naturally occurring mouse hepatitis virus. Am. J. Pathol. 106:110–117.
Bosma, G. C., R. P. Custer, and M. J. Bosma. 1983. A severe combined immunodeficiency mutation in the mouse. Nature (London) 301:527–530.
Bosma, G. C., M. Fried, R. P. Custer, A. Carroll, D. M. Gibson, and M. J. Bosma. 1988. Evidence of functional lymphocytes in some (leaky) scid mice. J. Exp. Med. 167:1016–1033.
Bosma, G. C., M. T. Davisson, N. R. Ruetsch, H. O. Sweet, L. D. Shultz, and M. J. Bosma. 1989. The mouse mutation severe combined immune deficiency (scid) is on chromosome 16. Immunogenetics 29:54–57.
Böttger, E. C., T. Hoffmann, U. Hadding, and D. Bitter-Suermann. 1985. Influence of genetically inherited complement deficiencies on humoral immune response in guinea pigs. J. Immunol. 135:4100–4107.
Böttger, E. C., T. Hoffmann, U. Hadding, and D. Bitter-Suermann. 1986a. Guinea pigs with

inherited deficiencies of complement components C2 or C4 have characteristics of immune complex disease. J. Clin. Invest. 78:689–695.

Böttger, E. C., S. Metzger, D. Bitter-Suermann, G. Stevenson, S. Kleindienst, and R. Burger. 1986b. Impaired humoral immune response in complement C3-deficient guinea pigs: Absence of secondary antibody response. Eur. J. Immunol. 16:1231–1235.

Bradley, D. J., B. A. Taylor, J. Blackwell, E. P. Evans, and J. Freeman. 1979. Regulation of *Leishmania* populations within the host. III. Mapping of the locus controlling susceptibility to visceral leishmaniasis in the mouse. Clin. Exp. Immunol. 37:7–14.

Bradley, L. M., and R. I. Mishell. 1981. Differential effects of glucocorticosteroids on the functions of helper and suppressor T lymphocytes. Proc. Natl. Acad. Sci. USA 78:3155–3159.

Brandt, E. J., R. T. Swank, and E. K. Novak. 1981. The murine Chediak-Higashi mutation and other murine pigmentation mutations. Pp. 99–117 in Immunologic Defects in Laboratory Animals, Vol. 1, M. E. Gershwin and B. Merchant, eds. New York: Plenum.

Breathnach, S. M., and S. I. Katz. 1986. Cell-mediated immunity in cutaneous disease. Human Pathol. 17:161–167.

Brick, J. E., S. E. Walker, and K. S. Wise. 1988. Hormone control of autoantibodies to calf thymus nuclear extract (CTE) and DNA in MRL-*lpr* and MRL- +/+ mice. Clin. Immunol. Immunpathol. 46:68–81.

Briles, D. E., R. M. Krause, and J. M. Davie. 1977. Immune response deficiency of BSVS mice. I. Identification of *Ir* gene differences between A/J and BSVS mice in the antistreptococcal group A carbohydrate response. Immunogenetics 4:381–392.

Briles, D. E., R. M. Perlmutter, D. Hansburg, J. R. Little, and J. M. Davie. 1979. Immune response deficiency of BSVS mice. II. Generalized deficiency to thymus-dependent antigens. Eur. J. Immunol. 9:255–261.

Briles, D. E., W. H. Benjamin, Jr., C. A. Williams, and J. M. Davie. 1981. A genetic locus responsible for salmonella susceptibility in BSVS mice is not responsible for the limited T-dependent immune responsiveness of BSVS mice. J. Immunol. 127:906–911.

Briles, D. E., M. Nahm, T. N. Marion, R. M. Perlmutter, and J. M. Davie. 1982. Streptococal group A carbohydrate has properties of both a thymus-independent (TI-2) and a thymus-dependent antigen. J. Immunol. 128:2032–2035.

Briles, D. E., W. Benjamin, Jr., B. Posey, S. M. Michalek, and J. R. McGhee. 1986. Independence of macrophage activation and expression of the alleles of the *Ity* (immunity to *typhimurium*) locus. Microb. Pathogen. 1:33–41.

Brinton, M. A., H. Arnheiter, and O. Haller. 1982. Interferon independence of genetically controlled resistance to flaviviruses. Infect. Immun. 36:284–288.

Brooke, H. C. 1926. Hairless mice. J. Hered. 17:173–174.

Brooks, C. G., P. J. Webb, R. A. Robins, G. Robinson, R. W. Baldwin, and M. F. W. Festing. 1980. Studies on the immunobiology of *rnu/rnu* 'nude' rats with congenital aplasia of the thymus. Eur. J. Immunol. 10:58–65.

Brown, A. M., and D. E. McFarlin. 1981. Relapsing experimental allergic encephalomyelitis in the SJL/J mouse. Lab. Invest. 45:278–284.

Brown, I. N., and A. A. Glynn. 1987. The *Ity/Lsh/Bcg* gene significantly affects mouse resistance to *Mycobacterium lepraemurium*. Immunology 62:587–591.

Brown, P. J., R. G. Bruce, D. F. Manson-Smith, and D. M. V. Parrott. 1981. Intestinal mast cell response in thymectomised and normal mice infected with *Trichinella spiralis*. Vet. Immunol. Immunopathol. 2:189–198.

Brownscheidle, C. M., V. Wootten, M. H. Mathieu, D. L. Davis, and I. A. Hofmann. 1983. The effects of maternal diabetes on fetal maturation and neonatal health. Metabolism 32(suppl. 1):148–155.

Brownstein, D. G. 1983. Genetics of natural resistance to Sendai virus infection in mice. Infect. Immun. 41:308–312.

Bruce, D. L. 1972. Halothane inhibition of phytohemagglutinin-induced transformation of lymphocytes. Anesthesiology 36:201–205.

Brunda, M. J., H. T. Holden, and R. B. Herberman. 1980. Augmentation of natural killer cell activity of beige mice by interferon and interferon inducers. Pp. 411–415 in Natural Cell-Mediated Immunity Against Tumors, R. B. Herberman, ed. New York: Academic Press.

Bubbers, J. E. 1983. Single-gene abrogation of spontaneous pleomorphic SJL/J mouse reticulum cell sarcoma expression. J. Natl. Cancer Inst. 71:795–799.

Bubbers, J. E. 1984. Identification and linkage analyses of a gene, *Rcs-1*, suppressing spontaneous SJL/J lymphoma expression. J. Natl. Cancer Inst. 72:441–446.

Burger, R., J. Gordon, G. Stevenson, G. Ramadori, B. Zanker, U. Hadding, and D. Bitter-Suermann. 1986. An inherited deficiency of the third component of complement, C3, in guinea pigs. Eur. J. Immunol. 16:7–11.

Buse, J. B., A. Ben-Nun, K. A. Klein, G. S. Eisenbarth, J. G. Seidman, and R. A. Jackson. 1984. Specific class II histocompatibility gene polymorphism in BB rats. Diabetes 33:700–703.

Butler, L., D. L. Guberski, and A. A. Like. 1988. Genetics of diabetes production in the Worcester colony of the BB rat. Pp. 74–78 in Frontiers in Diabetes Research. Lessons from Animals Diabetes II, E. Shafrir and A. E. Renold. London: John Libbey & Co.

Calich, V. L. G., E. Burger, S. S. Kashino, R. A. Fazioli, and L. M. Singer-Vermes. 1987. Resistance to *Paracoccidioides brasiliensis* in mice is controlled by a single dominant autosomal gene. Infect. Immun. 55:1919–1923.

Callisher, C. H., and W. P. Rowe. 1966. Mouse hepatitis, Reo-3, and the Theiler viruses. Pp. 67–75 in Viruses of Laboratory Rodents. National Cancer Institute Monograph No. 20. Bethesda, Md.: U.S. Department of Health, Education, and Welfare.

Calvano, S. E. 1986. Hormonal mediation of immune dysfunction following thermal and traumatic injury. Pp. 111–142 in Host Defenses in Trauma and Surgery, Vol. 6, Advances in Host Defense Mechanisms, J. M. Davis and G. T. Shires, eds. New York: Raven Press.

Canadian Council on Animal Care. 1980. Guide to the Care and Use of Experimental Animals, Vol. 1. Ottawa, Ontario: Canadian Council on Animal Care. 112 pp.

Cantor, H., L. McVay-Boudreau, J. Hugenberger, K. Naidof, F. W. Shen, and R. K. Gershon. 1978. Immunoregulatory circuits among T-cell sets. II. Physiologic role of feedback inhibition in vivo: Absence in NZB mice. J. Exp. Med. 147:1116–1125.

Caren, L. D., and L. T. Rosenberg. 1966. The role of complement in resistance to endogenous and exogenous infection with a common mouse pathogen, *Corynebacterium kutscheri*. J. Exp. Med. 124:689–699.

Carroll, A. M., and M. J. Bosma. In press. Detection and characterization of functional T cells in some (leaky) *scid* mice. Eur. J. Immunol.

Carter, T. C. 1954. Research news: New mutants. Mouse News Lett. 11:16.

Carthew, P., and S. Sparrow. 1980. Sendai virus in nude and germ-free rats. Res. Vet. Sci. 29:289–292.

Casarett, G. W. 1980. Radiation Histopathology, Vol. 1. Boca Raton, Fla.: CRC Press. 139 pp.

Cassell, G. H. 1982. The pathogenic potential of mycoplasmas: *Mycoplasma pulmonis* as a model. Rev. Infect. Dis. 4:S18-S34.

Castle, W. E., and H. D. King. 1944. Linkage studies of the rat (*Rattus norvegicus*). VI. Proc. Natl. Acad. Sci. USA 30:79–82.

Cerquetti, C. M., D. O. Sordelli, R. A. Ortegon, and J. A. Ballanti. 1983. Impaired lung defenses against *Staphylococcus aureus* in mice with hereditary deficiency of the fifth component of complement. Infect. Immun. 41:1071–1076.

Chambers, T., and J. Loutit. 1979. A functional assessment of macrophages from osteopetrotic mice. J. Pathol. 129:57–63.

Chandra, R. K., and B. Au. 1980. Spleen hemolytic plaque-forming cell response and generation of cytotoxic cells in genetically obese (C57BL/6J-*ob/ob*) mice. Int. Arch. Allergy Appl. Immunol. 62:94–98.

Chang, S.-S., and W. H. Hildemann. 1964. Inheritance of susceptibility to polyoma virus in mice. J. Natl. Cancer Inst. 33:303–313.

Chappel, C. I., and W. R. Chappel. 1983. The discovery and development of the BB rat colony: An animal model of spontaneous diabetes mellitus. Metabolism 32(suppl. 1):8–10.

Chase, H. B. 1959. Mutant stocks. Mouse News Lett. 21:21.

Chase, H. B. 1965. Change of symbol. Mouse News Lett. 33:17.

Chedid, L., M. Parant, C. Damais, F. Parant, D. Juy, and A. Galelli. 1976. Failure of endotoxin to increase nonspecific resistance to infection of lipopolysaccharide low-responder mice. Infect. Immun. 13:722–727.

Chen, J., and A. L. Goldstein. 1985. Thymosins and other thymic hormones. Pp. 121–140 in Biological Response Modifiers. New Approaches to Disease Intervention, P. F. Torrence, ed. Orlando, Fla.: Academic Press.

Chi, E. Y., E. Ignácio, and D. Lagunoff. 1978. Mast cell granule formation in the beige mouse. J. Histochem. Cytochem. 26:131–137.

Cinader, B., and S. Dubiski. 1964. Effect of autologous protein on the specificity of the antibody response: Mouse and rabbit antibody to MuB1. Nature (London) 202:102–103.

Claësson, M. H., N. Warner, and I. F. C. McKenzie. 1978. B-lymphocyte colony-forming cells in the SJL/J mouse thymus graft repopulation: Independence on host thymus and graft genotype. Scand. J. Immunol. 8:557–561.

Claman, H. N. 1972. Corticosteroids and lymphoid cells. N. Engl. J. Med. 287:388–397.

Clark, D. A., B. Cinader, and S. W. Koh. 1981. Age-dependent separation of class-specific suppressor cells in thymus of SJL/J mice. Immunol. 3:189–194.

Clark, E. A., L. D. Shultz, and S. B. Pollack. 1981. Mutations in mice that influence natural killer (NK) cell activity. Immunogenetics 12:601–613.

Clark, E. A., J. B. Roths, E. D. Murphy, J. A. Ledbetter, and J. A. Clagett. 1982. The beige (*bg*) gene influences the development of autoimmune disease in SB/Le male mice. Pp. 301–306 in NK Cells and Other Natural Effector Cells, R. B. Herberman, ed. New York: Academic Press.

Clatch, R. J., R. W. Melvold, S. D. Miller, and H. L. Lipton. 1985. Theiler's murine encephalomyelitis virus (TMEV)-induced demyelinating disease in mice is influenced by the H-2D region: Correlation with TMEV-specific delayed-type hypersensitivity. J. Immunol. 135:1408–1414.

Cockerell, G. L., E. A. Hoover, S. Krakowka, R. G. Olsen, and D. S. Yohn. 1976. Lymphocyte mitogen reactivity and enumeration of circulating B- and T-cells during feline leukemia virus infection in the cat. J. Natl. Cancer Inst. 57:1095–1099.

Cohen, D. I., A. D. Steinberg, W. E. Paul, and M. M. Davis. 1985. Expression of an X-linked gene family (XLR) in late-stage B cells and its alteration by the *xid* mutation. Nature (London) 314:372–374.

Cohen, P. L., I. Scher, and D. E. Mosier. 1976. In vitro studies of the genetically determined unresponsiveness to thymus-independent antigens in CBA/N mice. J. Immunol. 116:301–304.

Coleman, D. L. 1978. Obese and diabetes: Two mutant genes causing diabetes-obesity syndromes in mice. Diabetologia 14:141–148.

Colle, E., R. D. Guttmann, and T. Seemayer. 1981. Spontaneous diabetes mellitus syndrome

in the rat. I. Association with the major histocompatibility complex. J. Exp. Med. 154:1237–1242.
Colle, E., R. D. Guttmann, T. Seemayer, and F. Michel. 1983. Spontaneous diabetes mellitus syndrome in the rat. IV. Immunogenetic interactions of MHC and non-MHC components of the syndrome. Metabolism 32(suppl. 1):54–61.
Collins, F. M., and G. B. Mackaness. 1968. Delayed hypersensitivity and Arthus reactivity in relation to host resistance in salmonella-infected mice. J. Immunol. 101:830–845.
Collins, M. J., Jr., and J. C. Parker. 1972. Murine virus contaminants of leukemia viruses and transplantable tumors. J. Natl. Cancer Inst. 49:1139–1143.
Colten, H. R. 1983. Molecular biology and biosynthesis of the complement proteins. Pp. 397–406 in Progress in Immunology, Vol. 5, Y. Yamamura and T. Tada, eds. Tokyo: Academic Press.
Colten, H. R., and M. M. Frank. 1972. Biosynthesis of the second (C2) and fourth (C4) components of complement in vitro by tissues isolated from guinea-pigs with genetically determined C4 deficiency. Immunology 22:991–999.
Cook, J. A., E. O. Hoffman, and N. R. DiLuzio. 1975. Influence of lead and cadmium on the susceptibility of rats to bacterial challenge. Proc. Soc. Exp. Biol. Med. 150:741–747.
Cooke, A., and P. Hutchings. 1984. Defective regulation of erythrocyte autoantibodies in SJL mice. Immunology 51:489–492.
Costa, O., and J. C. Monier. 1986. Antihistone antibodies detected by micro-ELISA and immunoblotting in mice with lupus-like syndrome (MRL/l, MRL/n, PN, and NZB strains). Clin. Immunol. Immunopathol. 40:276–282.
Cotton, W. R., and J. F. Gaines. 1974. Unerupted dentition secondary to congenital osteopetrosis in the Osborne-Mendel rat. Proc. Soc. Exp. Biol. Med. 146:554–561.
Coutinho, A., G. Müller, and E. Gronowicz. 1975. Genetical control of B-cell responses. J. Exp. Med. 142:253–258.
Cowdery, J. S., S. M. Jacobi, A. K. Pitts, and T. L. Tyler. 1987. Defective B cell clonal regulation and autoantibody production in New Zealand black mice. J. Immunol. 138:760–764.
Cowing, C., D. Harris, C. Garabedian, M. Lukic, and S. Leskowitz. 1977. Immunologic tolerance to heterologous immunoglobulin: Its relation to *in vitro* filtration by macrophages. J. Immunol. 119:256–262.
Cowing, C., C. Garabedian, and S. Leskowitz. 1979. Strain differences in tolerance induction to human γ-globulin subclasses: Dependence on macrophages. Cell. Immunol. 47:407–415.
Cramer, P., and R. B. Painter. 1981. Bleomycin-resistant DNA synthesis in ataxia telangiectasia cells. Nature (London) 291:671–672.
Creighton, W. D., R. M. Zinkernagel, and F. J. Dixon. 1979. T cell-mediated immune responses of lupus-prone BXSB mice and other murine strains. Clin. Exp. Immunol. 37:181–189.
Crispens, C. G. 1973. Some characteristics of strain SJL/JDg mice. Lab. Anim. Sci. 23:408–413.
Crispens, C. G., Jr. 1979. Reproduction and growth characteristics: Mouse. Part I. Basic data. Pages 45–47 in Inbred and Genetically Defined Strains of Laboratory Animals. Part 1. Mouse and Rat, P. L. Altman and D. D. Katz, eds. Bethesda, Md.: Federation of American Societies for Experimental Biology.
Cross, S. S., H. C. Morse III, and R. Asofsky. 1976. Neonatal infection with mouse thymic virus. Differential effects on T cells mediating the graft-versus-host reaction. J. Immunol. 117:635–638.
Crowle, A. J., and M. May. 1978. A hanging drop macrophage function test. J. Reticuloendothel. Soc. 24:169–185.

Crowle, P. K. 1983. Mucosal mast cell reconstitution and *Nippostrongylus brasiliensis* rejection by W/W^v mice. J. Parasitol. 69:66–69.

Crowle, P. K., and N. D. Reed. 1981. Rejection of the intestinal parasite *Nippostrongylus brasiliensis* by mast cell-deficient W/W^v anemic mice. Infect. Immun. 33:54–58.

Cullen, B. F. 1974. The effect of halothane and nitrous oxide on phagocytosis and human leukocyte metabolism. Anesthesiol. Analges. 53:531–535.

Curtis, J., and J. L. Turk. 1984. Resistance to subcutaneous infection with *Mycobacterium lepraemurium* is controlled by more than one gene. Infect. Immun. 43:925–930.

Curtis, J., H. O. Adu, and J. L. Turk. 1982. *H-2* linkage control of resistance to subcutaneous infection with *Mycobacterium lepraemurium*. Infect. Immun. 38:434–439.

Custer, R. P., G. C. Bosma, and M. J. Bosma. 1985. Severe combined immunodeficiency (SCID) in the mouse. Am. J. Pathol. 120:464–477.

Czerski, P., E. Paprocka-Stonka, M. Siekierzyński, and A. Stolarska. 1974. Influence of microwave radiation on the hematopoietic system. Pp. 67–74 in Biologic Effects and Health Hazards of Microwave Radiation. Proceedings of an International Symposium held October 15–18, 1973, in Warsaw, Poland. Warsaw: Polish Medical Publishers.

Czitrom, A. A., S. Edwards, R. A. Phillips, M. J. Bosma, P. Marrack, and J. W. Kappler. 1985. The function of antigen-presenting cells in mice with severe combined immunodeficiency. J. Immunol. 134:2276–2280.

Dandoy, F., J. DeMaeyer-Guignard, D. Bailey, and E. DeMaeyer. 1982. Mouse genes influence antiviral action of interferon in vivo. Infect. Immun. 38:89–93.

D'Andrea, B. J., G. L. Wilson, and J. E. Craighead. 1981. Effect of genetic obesity in mice on the induction of diabetes by encephalomyocarditis virus. Diabetes 30:451–454.

Dang, H., and R. J. Harbeck. 1984. The *in vivo* and *in vitro* glomerular deposition of isolated anti-double-stranded-DNA antibodies in NZB/W mice. Clin. Immunol. Immunopathol. 30:265–278.

Dardenne, M., W. Savino, and J.-F. Bach. 1984. Autoimmune mice develop antibodies to thymic hormone: Production of anti-thymulin monoclonal autoantibodies from diabetic (*db/db*) and B/W mice. J. Immunol. 133:740–743.

Das, S., and S. Leskowitz. 1970. The kinetics of *in vivo* tolerance introduction in mice. J. Immunol. 105:938–943.

Das, S., and S. Leskowitz. 1974. The cellular basis for tolerance or immunity to bovine γ-globulin in mice. J. Immunol. 112:107–114.

Datta, S. K., and R. S. Schwartz. 1976. Genetics of expression of xenotropic virus and autoimmunity in NZB mice. Nature (London) 263:412–414.

Datta, S. K., and R. S. Schwartz. 1977. Mendelian segregation of loci controlling xenotropic virus production in NZB crosses. Virology 83:449–452.

Datta, S. K., N. Manny, C. Andrzejewski, J. André-Schwartz, and R. S. Schwartz. 1978a. Genetic studies of autoimmunity and retrovirus expression in crosses of New Zealand black mice. I. Xenotropic virus. J. Exp. Med. 147:854–871.

Datta, S. K., P. J. McConahey, N. Manny, A. N. Theofilopoulos, F. J. Dixon, and R. S. Schwartz. 1978b. Genetic studies of autoimmunity and retrovirus expression in crosses of New Zealand black mice. II. The viral envelope glycoprotein gp70. J. Exp. Med. 147:872–881.

Datta, S. K., F. L. Owen, J. E. Womack, and R. J. Riblet. 1982. Analysis of recombinant inbred lines derived from "autoimmune" (NZB) and "high leukemia" (C58) strains: Independent multigenic systems control B cell hyperactivity, retrovirus expression, and autoimmunity. J. Immunol. 129:1539–1544.

Datta, S. K., H. Patel, and D. Berry. 1987. Induction of a cationic shift in IgG anti-DNA autoantibodies. Role of T helper cells with classical and novel phenotypes in three murine models of lupus nephritis. J. Exp. Med. 165:1252–1268.

David, C. S. 1984. H-2 regulation of autoimmune and infectious diseases. Pp. 63–75 in Immunogenetics. Its Application to Clinical Medicine, T. Sasazuki and T. Tada, eds. Proceedings of the International Symposium on Immunogenetics, held August 17–19, 1983, in Tokyo, Japan. Tokyo: Academic Press.

Davidson, W. F. 1982. Immunologic abnormalities of the autoimmune mouse, Palmerston North. J. Immunol. 129:751–758.

Davidson, W. F., K. L. Holmes, J. B. Roths, and H. C. Morse III. 1985. Immunologic abnormalities of mice bearing the *gld* mutation suggest a common pathway for murine nonmalignant lymphoproliferative disorders with autoimmunity. Proc. Natl. Acad. Sci. USA 82:1219–1223.

Davidson, W. F., F. J. Dumont, H. G. Bedigian, B. J. Fowlkes, and H. C. Morse III. 1986. Phenotypic, functional, and molecular genetic comparisons of the abnormal lymphoid cells of C3H-*lpr/lpr* and C3H-*gld/gld* mice. J. Immunol. 136:4075–4084.

Davies, A. J. S. 1969. The thymus and the cellular basis of immunity. Transplant. Rev. 1:43–91.

Davies, E. V., A. M. T. Singleton, and J. M. Blackwell. 1988. Differences in *Lsh* gene control over systemic *Leishmania major* and *Leishmania donovani* or *Leishmania mexicana mexicana*. Infections are caused by differential targeting to infiltrating and resident liver macrophage populations. Infect. Immun. 56:1128–1134.

Davis, J. K., R. B. Thorp, and G. H. Cassell. 1982. Murine respiratory mycoplasmosis: Lesion development in relation to hereditary and immune responsiveness. Rev. Infect. Dis. 4:S239 (abstr.).

Debray-Sachs, M., M. Dardenne, P. Sai, W. Savino, M. C. Quiniou, D. Boillot, W. Gepts, and R. Assan. 1983. Anti-islet immunity and thymic dysfunction in the mutant diabetic C57BL/KsJ-*db/db* mouse. Diabetes 32:1048–1054.

De Cordoba, S. R., and P. Rubinstein. 1986. Quantitative variations of the C3b/C4b receptor (CR1) in human erythrocytes are controlled by genes within the regulator of complement activation (RCA) gene cluster. J. Exp. Med. 164:1274–1283.

De Maeyer-Guignard, J., A. Cachard-Thomas, and E. De Maeyer. 1984. Naturally occurring anti-interferon antibodies in LOU/C rats. J. Immunol. 133:775–778.

Denis, M., A. Forget, M. Pelletier, and E. Skamene. 1988. Pleiotropic effects of the *Bcg* gene. I. Antigen presentation in genetically susceptible and resistant congenic mouse strains. J. Immunol. 140:2395–2400.

Dent, P. B. 1972. Immunodepression by oncogenic viruses. Prog. Med. Virol. 14:1–35.

DeRossi, A., E. D'Andrea, G. Biasi, D. Collavo, and L. Chieco-Bianchi. 1983. Protection from spontaneous lymphoma development in SJL/J (v$^+$) mice neonatally injected with dualtropic SJL-151 virus. Proc. Natl. Acad. Sci. USA 80:2775–2779.

Descotes, J. 1988. Immunotoxicology of Drugs and Chemicals, 2d ed. Amsterdam: Elsevier. 444 pp.

Diamond, R. D., J. E. May, M. A. Kane, M. M. Frank, and J. E. Bennett. 1974. The role of the classical and alternate complement pathways in host defenses against *Cryptococcus neoformans* infection. J. Immunol. 112:2260–2270.

Dickie, M. M. 1964. Lethargic (*lh*) mutant mouse. Mouse News Lett. 30:31.

Dickie, M. M. 1967. New mutations. Mouse News Lett. 36:39.

Dinsley, M. 1976. Gnotobiotic animals: 2. Pp. 147–149 in The UFAW Handbook on the Care and Management of Laboratory Animals, 5th ed., Universities Federation for Animal Welfare (UFAW), ed. Edinburgh: Churchill Livingstone.

Dixon, F. J., M. B. A. Oldstone, and G. Tonietti. 1969. Virus-induced immune-complex-type glomerulonephritis. Transplant. Proc. 1:945–948.

Doody, D. P., E. J. Wilson, D. N. Medearis, and R. H. Rubin. 1986. Changes in the phenotype of T-cell subset determinants following murine cytomegalovirus infection. Clin. Immunol. Immunopathol. 40:466–475.

Dörries, R., R. Watanabe, H. Wege, and V. ter Meulen. 1987. Analysis of the intrathecal humoral immune response in Brown Norway (BN) rats, infected with the murine coronavirus JHM. J. Neuroimmunol. 14:305–316.

Dorshkind, K., G. M. Keller, R. A. Phillips, R. G. Miller, G. C. Bosma, M. O'Toole, and M. J. Bosma. 1984. Functional status of cells from lymphoid and myeloid tissues in mice with severe combined immunodeficiency disease. J. Immunol. 132:1804–1808.

Dorshkind, K., S. B. Pollack, M. J. Bosma, and R. A. Phillips. 1985. Natural killer (NK) cells are present in mice with severe combined immunodeficiency (*scid*). J. Immunol. 134:3798–3801.

Dueymes, M., G. J. Fournié, M. Mignon-Conté, S. In, and J. J. Conté. 1986. Prevention of lupus diseases in MRL/1, NZB × NZW, and BXSB mice treated with a cyclophosphazene derived drug. Clin. Immunol. Immunopathol. 41:193–205.

Dumont, F., F. Robert, and P. Bischoff. 1979. T and B lymphocytes in pituitary dwarf Snell-Bagg mice. Immunology 38:23–31.

Dung, H. C. 1981. Lethargic mice. Pp. 17–37 in Immunologic Defects in Laboratory Animals, Vol. 2, M. E. Gershwin and B. Merchant, eds. New York: Plenum.

Dunn, L. C. 1937. Studies on spotting patterns. II. Genetic analysis of variegated spotting in the house mouse. Genetics 22:43–64.

Dunn, T. B. 1954. Normal and pathologic anatomy of the reticular tissue in laboratory mice, with a classification and discussion of neoplasms. J. Natl. Cancer Inst. 14:1281–1433.

Dupuy, J.-M., C. Dupuy, and D. Décarie. 1984. Genetically determined resistance to mouse hepatitis virus 3 is expressed in hematopoietic donor cells in radiation chimeras. J. Immunol. 133:1609–1613.

Duquesnoy, R. J., and R. A. Good. 1970. Prevention of immunologic deficiency in pituitary dwarf mice by prolonged nursing. J. Immunol. 104:1553–1555.

Duquesnoy, R. J., and G. M. Pedersen. 1981. Immunologic and hematologic deficiencies of the hypopituitary dwarf mouse. Pp. 309–324 in Immunologic Defects in Laboratory Animals, Vol. 1, M. E. Gershwin and B. Merchant, eds. New York: Plenum.

Duquesnoy, R. J., P. K. Kalpaktsoglou, and R. A. Good. 1970. Immunological studies of the Snell-Bagg pituitary dwarf mouse. Proc. Soc. Exp. Biol. Med. 133:201–206.

Durant, S. 1986. *In vivo* effects of catecholamines and glucocorticoids on mouse thymic cAMP content and thymolysis. Cell. Immunol. 102:136–143.

Dyrberg, T., A. F. Nakhooda, S. Baekkeskov, A. Lernmark, P. Poussier, and E. B. Marliss. 1982. Islet cell surface antibodies and lymphocyte antibodies in the spontaneously diabetic BB Wistar rat. Diabetes 31:278–281.

Easmon, C. S. F., and A. A. Glynn. 1976. Comparison of subcutaneous and intraperitoneal staphylococcal infections in normal and complement- deficient mice. Infect. Immun. 13:399–406.

East, J. 1970. Immunopathology and neoplasms in New Zealand black (NZB) and SJL/J mice. Prog. Exp. Tumor Res. 13:85–134.

East, J., A. B. de Sousa, P. R. Prosser, and H. Jaquet. 1967. Malignant changes in New Zealand black mice. Clin. Exp. Immunol. 2:427–443.

Eastcott, J. W., R. S. Schwartz, and S. K. Datta. 1983. Genetic analysis of the inheritance of B cell hyperactivity in relation to the development of autoantibodies and glomerulonephritis in NZB × SWR crosses. J. Immunol. 131:2232–2239.

Eaton, G. J. 1976. Hair growth cycles and wave patterns in "nude" mice. Transplantation 22:217–222.

Eaton, G. J., H. C. Outzen, R. P. Custer, and F. N. Johnson. 1975. Husbandry of the "nude" mouse in conventional and germfree environments. Lab. Anim. Sci. 25:309–314.

Edén, C. S., R. Shahin, and D. Briles. 1988. Host resistance to mucosal gram-negative infection. Susceptibility of lipopolysaccharide nonresponder mice. J. Immunol. 140:3180–3185.
Eicher, E. M. 1976. Remutations. Mouse News Lett. 54:40.
Eicher, E. M., and W. G. Beamer. 1980. New mouse *dw* allele: Genetic location and effects on lifespan and growth hormone levels. J. Hered. 71:187–190.
Eilat, D., D. M. Webster, and A. R. Rees. 1988. V region sequences of anti-DNA and anti-RNA autoantibodies from NZB/NZW F_1 mice. J. Immunol. 141:1745–1753.
Eisenberg, R. A., and F. J. Dixon. 1980. Effect of castration on male-determined acceleration of autoimmune disease in BXSB mice. J. Immunol. 125:1959–1961.
Ek-Rylander, B., S. C. Marks, Jr., L. Hammarstrom, and G. Andersson. 1989. Osteoclastic acid ATPase—Biochemical and histochemical studies of the three osteopetrotic mutations in the rat. Bone Mineral 5:309–321.
Elder, M. E., and N. K. MacLaren. 1983. Identification of profound peripheral T lymphocyte immunodeficiencies in the spontaneously diabetic BB rat. J. Immunol. 130:1723–1731.
Elder, M., N. MacLaren, W. Riley, and T. McConnell. 1982. Gastric parietal cell and other autoantibodies in the BB rat. Diabetes 31:313–318.
Elftman, H., and O. Wegelius. 1959. Anterior pituitary cytology of the dwarf mouse. Anat. Rec. 135.43–49.
Elin, R. J., J. B. Edelin, and S. M. Wolff. 1974. Infection and immunoglobulin concentrations in Chediak-Higashi mice. Infect. Immun. 10:88–91.
Elliott, R. B., S. N. Reddy, N. J. Bibby, and K. Kida. 1988. Dietary prevention of diabetes in the non-obese diabetic mouse. Diabetologia 31:62–64.
Ellman, L., I. Green, and M. Frank. 1970. Genetically controlled total deficiency of the fourth component of complement in the guinea pig. Science 170:74–75.
Ellman, L., I. Green, F. Judge, and M. M. Frank. 1971. In vivo studies in C4-deficient guinea pigs. J. Exp. Med. 134:162–175.
Engleman, E. G., G. Sonnenfeld, M. Dauphinee, J. S. Greenspan, N. Talal, H. O. McDevitt, and T. C. Merigan. 1981. Treatment of NZB/NZW F_1 hybrid mice with *Mycobacterium bovis* strain BCG or type II interferon preparations accelerates autoimmune disease. Arthritis Rheum. 24:1396–1402.
Erickson, K. L., and M. E. Gershwin. 1981. Hereditarily athymic-asplenic (lasat) mice. Pp. 297–308 in Immunologic Defects in Laboratory Animals, Vol. 1, M. E. Gershwin and B. Merchant, eds. New York: Plenum.
Erickson, R. P., D. K. Tachibana, L. A. Herzenberg, and L. T. Rosenberg. 1964. A single gene controlling hemolytic complement and a serum antigen in the mouse. J. Immunol. 92:611–615.
Evans, M. M., W. G. Williamson, and W. J. Irvine. 1968. The appearance of immunological competence at an early age in New Zealand black mice. Clin. Exp. Immunol. 3:375–383.
Evans, S. F., N. W. Nisbet, M. J. Marshall, and D. Catty. 1985. A secondary immune deficiency in the Fatty/Orl-*op* rat. Br. J. Exp. Pathol. 66:251–255.
Falconer, D. S., and J. H. Isaacson. 1959. Adipose, a new inherited obesity of the mouse. J. Hered. 50:290–292.
Fenner, F. 1949. Mouse-pox (infectious ectromelia of mice): A review. J. Immunol. 63:341–373.
Fernandes, G. 1984. Nutritional factors: Modulating effects on immune function and aging. Pharmacol. Rev. 36:123S–129S.
Fernandes, G., E. J. Yunis, and R. A. Good. 1975. Depression of cytotoxic T cell subpopulation in mice by hydrocortisone treatment. Clin. Immunol. Immunopathol. 4:304–313.
Fernandes, G., E. J. Yunis, and R. A. Good. 1976. Influence of diet on survival of mice. Proc. Natl. Acad. Sci. USA 73:1279–1283.

Fernandes, G., B. S. Handwerger, E. J. Yunis, and D. M. Brown. 1978. Immune response in the mutant diabetic C57BL/Ks-db+ mouse. Discrepancies between in vitro and in vivo immunological assays. J. Clin. Invest. 61:243–250.

Ferreira, A., M. Takahashi, and V. Nussenzweig. 1977. Purification and characterization of mouse serum protein with specific binding affinity for C4 (Ss protein). J. Exp. Med. 146:1001–1018.

Festing, M. F. W. 1981. Athymic nude rats. Pp. 267–283 in Immunologic Defects in Laboratory Animals, Vol. 1, M. E. Gershwin and B. Merchant, eds. New York: Plenum.

Festing, M. F. W., D. May, T. A. Connors, D. Lovell, and S. Sparrow. 1978. An athymic nude mutation in the rat. Nature (London) 274:365–366.

Fidler, I. J. 1977. Depression of macrophages in mice drinking hyperchlorinated water. Nature (London) 270:735–736.

Fieser, T. M., M. E. Gershwin, A. D. Steinberg, F. J. Dixon, and A. N. Theofilopoulos. 1984. Abrogation of murine lupus by the *xid* gene is associated with reduced responsiveness of B cells to T-cell-helper signals. Cell. Immunol. 87:708–713.

Fitzgerald, K. L., and N. M. Ponzio. 1981. Natural killer cell activity in reticulum-cell sarcomas of SJL/J mice. II. Analysis of RCS-associated NK activity. Int. J. Cancer 28:627–634.

Flanagan, S. P. 1966. "Nude," a new hairless gene with pleiotropic effects in the mouse. Genet. Res. 8:295–309.

Flatt, P. R., A. J. Bone, and C. J. Bailey. 1985. Islet cell surface antibodies in genetically obese hyperglycaemic (*ob/ob*) mice. Biosci. Rep. 5:715–720.

Fletcher, M. P., R. M. Ikeda, and M. E. Gershwin. 1977. Splenic influence of T cell function: The immunobiology of the inbred hereditarily asplenic mouse. J. Immunol. 119:110–117.

Fodstad, Ø., C. T. Hansen, G. B. Cannon, and M. R. Boyd. 1984a. Immune characteristics of the beige-nude mouse. A model for studying immune surveillance. Scand. J. Immunol. 20:267–272.

Fodstad, Ø., C. T. Hansen, G. B. Cannon, C. N. Statham, G. R. Lichtenstein, and M. R. Boyd. 1984b. Lack of correlation between natural killer activity and tumor growth control in nude mice with different immune defects. Cancer Res. 44:4403–4408.

Fossum, S., M. E. Smith, E. B. Bell, and W. L. Ford. 1980. The architecture of rat lymph nodes. III. The lymph nodes and lymph-borne cells of the congenitally athymic nude rat (*rnu*). Scand. J. Immunol. 12:421–432.

Foster, H. L., J. D. Small, and J. G. Fox, eds. 1982. The Mouse in Biomedical Research. Vol. II: Diseases. New York: Academic Press. 449 pp.

Fox, J. G., B. J. Cohen, and F. M. Loew, eds. 1984. Laboratory Animal Medicine. Orlando, Fla.: Academic Press. 750 pp.

Fraker, P. J., S. M. Haas, and R. W. Luecke. 1977. Effect of zinc deficiency on the immune response of the young adult A/J mouse. J. Nutr. 107:1889–1895.

Frank, M. M., J. May, T. Gaither, and L. Ellman. 1971. In vitro studies of complement function in sera of C4-deficient guinea pigs. J. Exp. Med. 134:176–187.

Fujiwara, M., and B. Cinader. 1974a. Cellular aspects of tolerance. IV. Strain variations of tolerance induceability. Cell. Immunol. 12:11–29.

Fujiwara, M., and B. Cinader. 1974b. Cellular aspects of tolerance. V. The *in vivo* cooperative role of accessory and thymus derived cells in responsiveness and unresponsiveness of SJL mice. Cell. Immunol. 12:194–204.

Fulop, G. M., and R. A. Phillips. 1986. Full reconstitution of the immune deficiency in *scid* with normal stem cells requires low-dose irradiation of the recipients. J. Immunol. 136:4438–4443.

Fulop, G. M., G. C. Bosma, M. J. Bosma, and R. A. Phillips. 1988. Early B-cell precursors in *scid* mice: Normal numbers of cells transformable with Abelson murine leukemia virus (A-MuLV). Cell. Immunol. 113:192–201.

Fultz, P. N., J. A. Shadduck, C. Y. Kang, and J. W. Streilein. 1981. Involvement of cells of hematopoietic origin in genetically determined resistance of Syrian hamsters to vesicular stomatitis virus. Infect. Immun. 34:540–549.

Gabrielsen, A. E., and R. A. Good. 1967. Chemical suppression of adaptive immunity. Adv. Immunol. 6:91–229.

Gallin, J. I., J. S. Bujak, E. Patten, and S. M. Wolff. 1974. Granulocyte function in the Chediak-Higashi syndrome of mice. Blood 43:201–206.

Garlinghouse, L. E., Jr., and G. L. Van Hoosier, Jr. 1978. Studies on adjuvant-induced arthritis, tumor transplantability, and serologic response to bovine serum albumin in Sendai virus infected rats. Am. J. Vet. Res. 39:297–300.

Gavalchin, J., and S. K. Datta. 1987. The NZB × SWR model of lupus nephritis. II. Autoantibodies deposited in renal lesions show a distinctive and restricted idiotypic diversity. J. Immunol. 138:138–148.

Gavalchin, J., J. A. Nicklas, J. W. Eastcott, M. P. Madaio, B. D. Stollar, R. S. Schwartz, and S. K. Datta. 1985. Lupus prone (SWR × NZB)F_1 mice produce potentially nephritogenic autoantibodies inherited from the normal SWR parent. J. Immunol. 134:885–894.

Gavalchin, J., R. A. Seder, and S. K. Datta. 1987. The NZB × SWR model of lupus nephritis. I. Cross-reactive idiotypes of monoclonal anti-DNA antibodies in relation to antigenic specificity, charge, and allotype. Identification of interconnected idiotype families inherited from the normal SWR and the autoimmune NZB parents. J. Immunol. 138:128–137.

Gazdar, A. F., W. Beitzel, and N. Talal. 1971. The age related response of New Zealand mice to a murine sarcoma virus. Clin. Exp. Immunol. 8:501–509.

Geiger, J. D., and J. I. Nagy. 1986. Lack of adenosine deaminase deficiency in the mutant mouse *wasted*. FEBS Lett. 208:431–434.

Gelfand, J. A., D. L. Hurley, A. S. Fauci, and M. M. Frank. 1978. Role of complement in host defense against experimental disseminated candidiasis. J. Infect. Dis. 138:9–16.

Gershwin, M. E., Y. Ohsugi, A. Ahmed, J. J. Castles, R. Scibienski, and R. M. Ikeda. 1980. Studies of congenitally immunologically mutant New Zealand mice. IV. Development of autoimmunity in congenitally athymic (nude) New Zealand black × white F1 hybrid mice. J. Immunol. 125:1189–1195.

Ghaffer, A., R. D. Paul, W. Lichter, L. L. Wellham, and M. M. Sigel. 1981. Selective action of alkylating agents on helper and suppressor functions. Pp. 210–228 in Biological Relevance of Immune Suppression as Induced by Genetic, Therapeutic and Environmental Factors, J. H. Dean and M. Padarathsingh, eds. Proceedings of a technology assessment workshop held November 13–14, 1979, in Williamsburg, Va. New York: Van Nostrand Reinhold.

Ghatak, S., K. Sainis, F. L. Owen, and S. K. Datta. 1987. T-cell-receptor β- and I-Aβ-chain genes of normal SWR mice are linked with the development of lupus nephritis in NZB × SWR crosses. Proc. Natl. Acad. Sci. USA 84:6850–6853.

Glode, L. M., and D. L. Rosenstreich. 1976. Genetic control of B cell activation by bacterial lipopolysaccharide is mediated by multiple distinct genes or alleles. J. Immunol. 117:2061–2066.

Glode, L. M., A. Jacques, S. E. Mergenhagen, and D. L. Rosenstreich. 1977. Resistance of macrophages from C3H/HeJ mice to the *in vitro* cytotoxic effects of endotoxin. J. Immunol. 119:162–166.

Glynn, A. A., and F. A. Medhurst. 1967. Possible extracellular and intracellular bactericidal actions of mouse complement. Nature (London) 213:608–610.

Goldberger, G., F. S. Cole, L. P. Einstein, H. S. Auerbach, D. Bitter-Suermann, and H. R. Colten. 1982. Biosynthesis of a structurally abnormal C2 complement protein by macrophages from C2-deficient guinea pigs. J. Immunol. 129:2061–2065.

Golding, B., H. Golding, P. G. Foiles, and J. I. Morton. 1983. CBA/N X-linked defect delays expression of the Y-linked accelerated autoimmune disease in BXSB mice. J. Immunol. 130:1043–1046.

Goldings, E. A. 1988. Defective B-cell tolerance in New Zealand black mice. Cell. Immunol. 113:183–191.

Goldings, E. A., P. L. Cohen, S. F. McFadden, M. Ziff, and E. S. Vitetta. 1980. Defective B cell tolerance in adult (NZB × NZW)F1. J. Exp. Med. 152:730–735.

Goldman, M. B., and J. N. Goldman. 1976. Relationship of functional levels of early components of complement to the H-2 complex of mice. J. Immunol. 117:1584–1588.

Goldner-Sauve, A. J., S. Ono, G. J. Prud'homme, R. D. Guttman, E. Colle, and A. Fuks. 1986. A permissive role for the major histocompatibility complex in the spontaneous syndrome of insulin dependent diabetes in the BB rat. Rat News Lett. 17:17–18.

Goldowitz, D., P. M. Shipman, J. F. Porter, and R. R. Schmidt. 1985. Longitudinal assessment of immunologic abnormalities of mice with the autosomal recessive mutation, *"wasted."* J. Immunol. 135:1806–1812.

Gordon, J. 1977. Modification of pigmentation patterns in allophenic mice by the *W* gene. Differentiation 9:19–27.

Green, E. L. 1966. Breeding systems. Pp. 11–22 in Biology of the Laboratory Mouse, 2d ed., E. L. Green, ed. New York: McGraw-Hill.

Green, M. C. 1966. Mutant genes and linkages. Pp. 87–150 in Biology of the Laboratory Mouse, 2d ed., E. L. Green, ed. New York: McGraw-Hill.

Green, M. C. 1967. A defect of the splanchnic mesoderm caused by the mutant gene dominant hemimelia in the mouse. Dev. Biol. 15:62–89.

Green, M. C. 1981. Catalog of mutant genes and polymorphic loci. Pp. 8–278 in Genetic Variants and Strains of the Laboratory Mouse, M. C. Green, ed. Stuttgart: Gustav Fischer Verlag.

Green, M. C., and L. D. Shultz. 1975. Motheaten, an immunodeficient mutant of the mouse. I. Genetics and pathology. J. Hered. 66:250–258.

Greenhouse, D. D., M. F. W. Festing, S. Hasan, and A. L. Cohen. In press. Catalog of inbred strains. In Genetic Monitoring of Inbred Strains of Rats. A Manual on Colony Management, Basic Monitoring Techniques, and Genetic Variants of the Laboratory Rat, H. J. Hedrich, ed. Stuttgart: Gustav Fischer Verlag.

Greenlee, J. E. 1981. Effect of host age on experimental K virus infection in mice. Infect. Immun. 33:297–303.

Greep, R. O. 1941. An hereditary absence of the incisor teeth. J. Hered. 32:397–398.

Greiner, D. L. 1986. The spontaneously diabetic Bio-Breeding (BB) rat: T cell subset deficiencies and their possible relationship to diabetes. Rat News Lett. 17:11–16.

Greiner, D. L., I. Goldschneider, K. L. Komschlies, E. S. Medlock, F. J. Bollum, and L. Shultz. 1986a. Defective lymphopoiesis in bone marrow of motheaten (*me/me*) and viable motheaten (*me^v/me^v*) mutant mice. I. Analysis of development of prothymocytes, early B lineage cells, and terminal deoxynucleotidyl transferase-positive cells. J. Exp. Med. 164:1129–1144.

Greiner, D. L., E. S. Handler, K. Nakano, J. P. Mordes, and A. A. Rossini. 1986b. Absence of the RT-6 T cell subset in diabetes-prone BB/W rats. J. Immunol. 136:148–151.

Gripois, D., C. Klein, and M. Valens. 1980. Influence of hypo- and hyperthyroidism on noradrenaline metabolism in brown adipose tissue of the developing rat. Biol. Neonate 37:53–59.

Gros, P., E. Skamene, and A. Forget. 1981. Genetic control of natural resistance to *Mycobacterium bovis* (BCG) in mice. J. Immunol. 127:2417–2421.

Groscurth, P., M. Müntener, and G. Töndury. 1974. The postnatal development of the thymus in the nude mouse. 1. Light microscopic observations. Pp. 31–36 in Proceedings of the First International Workshop on Nude Mice, J. Rygaard and C. O. Povlsen, eds. Proceedings of a workshop held October 11–13, 1973, in Scanticon, Aarhus, Denmark. Stuttgart: Gustav Fischer Verlag.

Grüneberg, H. 1935. A new sub-lethal colour mutation in the house mouse. Proc. R. Soc. Ser. B 118:321–342.

Grüneberg, H. 1936. Grey-lethal, a new mutation in the house mouse. J. Hered. 27:105–109.

Grüneberg, H. 1938. Some new data on the grey-lethal mouse. J. Genet. 36:153–170.

Guberski, D. L., L. Butler, and A. A. Like. 1988. The BBZ/Wor rat: An obese animal with autoimmune diabetes. Pp. 268–271 in Frontiers in Diabetes Research. Lessons from Animal Diabetes II, E. Shafrir and A. E. Renold, eds. London: John Libbey & Co.

Guberski, D. L., L. Butler, W. Kastern, and A. A. Like. In press. Genetic studies in inbred BB/Wor rats: Analysis of progeny produced by crossing lymphopenic diabetes-prone with non lymphopenic diabetic rats. Diabetes.

Guttmann, R. D., E. Colle, F. Michel, and T. Seemayer. 1983. Spontaneous diabetes mellitus in the rat. II. T lymphopenia and its association with clinical disease and pancreatic lymphocytic infiltration. J. Immunol. 130:1732–1735.

Ha, T. Y., N. D. Reed, and P. K. Crowle. 1983. Delayed expulsion of adult *Trichinella spiralis* by mast cell-deficient W/W^v mice. Infect. Immun. 41:445–447.

Hackett, J., G. C. Bosma, M. J. Bosma, M. Bennett, and V. Kumar. 1986. Transplantable progenitors of natural killer cells are distinct from those of T and B lymphocytes. Proc. Natl. Acad. Sci. USA 83:3427–3431.

Haddada, H., I. Chouroulinkov, and C. De Vaux St Cyr. 1982. Nude Syrian hamsters: Some immunological and histological characteristics. Immunol. Lett. 4:327–333.

Hadley, J. G., D. E. Gardner, D. L. Coffin, and D. B. Menzel. 1977. Inhibition of antibody-mediated rosette formation by alveolar macrophages: A sensitive assay for metal toxicity. J. Reticuloendothel. Soc. 22:417–425.

Hall, R. E., and H. R. Colten. 1977. Cell-free synthesis of the fourth component of guinea pig complement (C4): Identification of a precursor of serum C4 (pro-C4). Proc. Natl. Acad. Sci. USA 74:1707–1710.

Halle-Pannenko, O., and M. Bruley-Rosset. 1985. Decreased graft-versus-host reaction and T cell cytolytic potential of beige mice. Transplantation 39:85–87.

Hamilton, J. R., J. C. Overall, Jr., and L. A. Glasgow. 1976. Synergistic effect on mortality in mice with murine cytomegalovirus and *Pseudomonas aeruginosa*, *Staphylococcus aureus*, or *Candida albicans* infections. Infect. Immun. 14:982–989.

Hammer, C. H., T. Gaither, and M. M. Frank. 1981. Complement deficiencies of laboratory animals. Pp. 207–240 in Immunologic Defects in Laboratory Animals, Vol. 2, M. E. Gershwin and B. Merchant, eds. New York: Plenum.

Hang, L., A. N. Theofilopoulos, and F. J. Dixon. 1982. A spontaneous rheumatoid arthritis-like disease in MRL/l mice. J. Exp. Med. 155:1690–1701.

Haran-Ghera, N., M. Ben-Yaakov, A. Peled, and Z. Bentwich. 1973. Immune status of SJL/J mice in relation to age and spontaneous tumor development. J. Natl. Cancer Inst. 50:1227–1235.

Harrison, D. E., and C. M. Astle. 1976. Population of lymphoid tissues in cured *W*-anemic mice by donor cells. Transplantation 22:42–46.

Harrison, D. E., J. R. Archer, and C. M. Astle. 1984. Effects of food restriction on aging: Separation of food intake and adiposity. Proc. Natl. Acad. Sci. USA 81:1835–1838.

Hashimoto, K., Y. Okada, T. Tajiri, H. Amano, N. Aoki, and S. Sasaki. 1973. Intestinal resistance in the experimental enteric infection of mice with a mouse adenovirus. III. Suppressive effect of cyclophosphamide on the establishment and duration of the intestinal resistance. Jpn. J. Microbiol. 17:503–511.

Haspel, M. V., T. Onodera, B. S. Prabnakar, M. Horita, H. Suzuki, and A. L. Notkins.

1983. Virus-induced autoimmunity: Monoclonal antibodies that react with endocrine tissues. Science 220:304–306.

Hass, G. M., G. H. Laing, R. M. Galt, and P. A. McCreary. 1981. Recent advances: Immunopathology of magnesium deficiency in rats. Induction of tumors; incidence, transmission and prevention of lymphoma-leukemia. Magnesium Bull. 3(suppl. 1a):217–228.

Hattori, M., J. B. Buse, R. A. Jackson, L. Glimcher, M. E. Dorf, M. Minami, S. Makino, K. Moriwaki, H. Kuzuya, H. Imura, W. M. Strauss, J. G. Seidman, and G. S. Eisenbarth. 1986. The NOD mouse: Recessive diabetogenic gene in the major histocompatibility complex. Science 231:733–735.

Hauser, W. E., Jr., and J. S. Remington. 1982. Effects of antibiotics on the immune response. Am. J. Med. 72:711–716.

Hayakawa, K., R. R. Hardy, D. R. Parks, and L. A. Herzenberg. 1983. The "Ly-1 B" cell subpopulation in normal, immunodefective, and autoimmune mice. J. Exp. Med. 157:202–218.

Hayashi, S.-I., P. L. Witte, L. D. Shultz, and P. W. Kincade. 1988. Lymphohemopoiesis in culture is prevented by interaction with adherent bone marrow cells from mutant viable motheaten mice. J. Immunol. 140:2139–2147.

Hedrich, H. J., K. Wonigeit, and R. Schwinzer. 1987. In vivo alloreactivity and xenoreactivity of athymic nude rats. Transplant. Proc. 19:3199–3202.

Heiniger, H. J., H. Meier, N. Kaliss, M. Cherry, H. W. Chen, and R. D. Stoner. 1974. Hereditary immunodeficiency and leukemogenesis in HRS/J mice. Cancer Res. 34:201–211.

Hemphill, E. F., M. L. Kasberk, and W. B. Buck. 1971. Lead suppression of mouse resistance to *Salmonella typhimurium*. Science 172:1031–1032.

Hendrickson, E. A., D. G. Schatz, and D. T. Weaver. 1988. The *scid* gene encodes a *trans*-acting factor that mediates the rejoining event of Ig gene rearrangement. Genes Dev. 2:817–829.

Hermann, L. M., W. J. White, and C. M. Lang. 1982. Prolonged exposure to acid, chlorine, or tetracycline in drinking water: Effects on delayed-type hypersensitivity, hemagglutination titers, and reticuloendothelial clearance rates in mice. Lab. Anim. Sci. 32:603–608.

Herold, K. C., W. Kastern, H. Markholst, A. Lernmark, and B. E. Andreasen. In press. Derivation of non-lymphopenic BB rats with an intercross breeding. Autoimmunity.

Hertwig, P. 1942. Neue mutationen und koppelungsgruppen bei der Hausmaus. Z. Indukt. Abstamm. Vererbungsl. 80:220–246.

Hetherington, C. M., and M. A. Hegan. 1975. Breeding nude (*nu/nu*) mice. Lab. Anim. (London) 9:19–20.

Hiai, H., H. Ikeda, H. Kaneshima, T. Odaka, and Y. Nishizuka. 1982. Slvr-1: A new epistatic host gene restricting expression of endogenous ecotropic virus in SL/Ni mice is distinct from Fv-4. Nippon Gan Gakkai Sokai Kiji (Proc. Annu. Meeting Jpn. Cancer Assoc.) 41:335 (abstr. 1175).

Hicks, J. T., F. A. Ennis, E. Kim, and M. Verbonitz. 1978. The importance of an intact complement pathway in recovery from a primary viral infection: Influenza in decomplemented and C5-deficient mice. J. Immunol. 121:1437–1445.

Higuchi, K., T. Yonezu, K. Kogishi, A. Matsumura, S. Takeshita, K. Higuchi, A. Kohno, M. Matsushita, M. Hosokawa, and T. Takeda. 1986. Purification and characterization of a senile amyloid-related antigenic substance (apoSAS$_{SAM}$) from mouse serum. ApoSAS$_{SAM}$ is an apoA-II apolipoprotein of mouse high density lipoproteins. J. Biol. Chem. 261:12834–12840.

Hobart, M. 1984. Agreement now the norm. The biochemistry and genetics of complement. Immunol. Today 5:121–125.

Hoiseth, S. K., and B. A. D. Stocker. 1981. Aromatic-dependent *Salmonella typhimurium* are non-virulent and effective as live vaccines. Nature (London) 291:238–239.

Hormaeche, C. E. 1980. The in vivo division and death rates of *Salmonella typhimurium* in the spleens of naturally resistant and susceptible mice measured by the superinfecting phage technique of Meynell. Immunology 41:973–979.

Hosokawa, T., M. Hosono, K. Higuchi, A. Aoike, K. Kawai, and T. Takeda. 1987a. Immune responses in newly developed short-lived SAM mice. I. Age-associated early decline in immune activities of cultured spleen cells. Immunology 62:419–423.

Hosokawa, T., M. Hosono, K. Hanada, A. Aoike, K. Kawai, and T. Takeda. 1987b. Immune responses in newly developed short-lived SAM mice. II. Selectively impaired T-helper cell activity in in vitro antibody response. Immunology 62:425–429.

Hotchin, J. 1962. The biology of lymphocytic choriomeningitis infection: Virus-induced immune disease. Cold Spring Harbor Symp. Quant. Biol. 27:479–499.

Howie, J. B., and L. O. Simpson. 1976. Autoimmune diseases in NZB mice and their hybrids. Pp. 124–141 (chapter 4, suppl. 2) in Lupus Erythematosus, A Review of the Current Status of Discoid and Systemic Lupus Erythematosus and Their Varients, revised 2d ed., E. L. Dubois, ed. Los Angeles: University of Southern California Press.

Hsu, C.-K. 1976. Immunology. Pp. 479–518 in Handbook of Laboratory Animal Science, Vol. III, E. C. Melby, Jr., and N. H. Altman, eds. Cleveland: CRC Press

Hsu, C.-K., A. E. New, and J. G. Mayo. 1980. Quality assurance of rodent models. Pp. 17–28 in Animal Quality and Models in Biomedical Research, A. Spiegel, S. Erichsen, and H. A. Solleveld, eds. Proceedings of the 7th Symposium of the International Council for Laboratory Animal Science (ICLAS) held August 21–23, 1979, in Utrecht, The Netherlands. Stuttgart: Gustav Fischer Verlag.

Huber, B. 1982. B cell differentiation antigens as probes for functional B cell subsets. Immunol. Rev. 64:57–79.

Huber, B., R. K. Gershon, and H. Cantor. 1977. Identification of a B-cell surface structure involved in antigen-dependent triggering: Absence of this structure on B cells from CBA/N mutant mice. J. Exp. Med. 145:10–20.

Hudson, L., and F. C. Hay. 1976. Radioimmunoassay. Pp. 230–232 in Practical Immunology. Oxford: Blackwell Scientific.

Hughes, W. T., D. L. Bartley, and B. M. Smith. 1983. A natural source of infection due to *Pneumocystis carinii*. J. Infect. Dis. 147:595 (summary).

Hummel, K. P., M. M. Dickie, and D. L. Coleman. 1966. Diabetes, a new mutation in the mouse. Science 153:1127–1128.

Hummel, K. P., D. L. Coleman, and P. W. Lane. 1972. The influence of genetic background on expression of mutations at the diabetes locus in the mouse. I. C57BL/KsJ and C57BL/6J strains. Biochem. Genet. 7:1–13.

Hummel, K. P., D. R. Chapman, and D. L. Coleman. 1973. Allele test. Mouse News Lett. 48:34.

Hunig, T., and M. J. Bevan. 1980. Specificity of cytotoxic T cells from athymic mice. J. Exp. Med. 152:688–702.

Hyde, R. R. 1923. Complement-deficient guinea pig serum. J. Immunol. 8:267–286.

ICLAS (International Council for Laboratory Animal Science). 1987. Interrelationships between the animal's nutrition and the experimental results. Pp. 1–21 in ICLAS Guidelines on the Selection and Formulation of Diets for Animals in Biomedical Research, M. E. Coates, ed. A report of the ICLAS Nutrition Working Group. London: Institute of Biology.

Ikeda, R. M., and M. E. Gershwin. 1978. Congenitally athymic (nude) and congenitally athymic-asplenic mice: Contrasts and comparisons. Pp. 201–210 in Animal Models of Comparative and Developmental Aspects of Immunity and Disease, M. E. Gershwin and E. L. Cooper, eds. New York: Pergamon.

Ikegami, H., S. Makino, M. Harada, G. S. Eisenbarth, and M. Hattori. 1988. The cataract Shionogi mouse, a sister strain of the non-obese diabetic mouse: Similar class II but different class I gene products. Diabetologia 31:254–258.

Ikehara, S., T. Nakamura, K. Sekita, E. Muso, R. Yasumizu, H. Ohtsuki, Y. Hamashima, and R. A. Good. 1987. Treatment of systemic and organ-specific autoimmune disease in mice by allogeneic bone marrow transplantation. Pp. 131–146 in Animal Models: Assessing the Scope of Their Use in Biomedical Research, J. Kawamata and E. C. Melby, Jr., eds. New York: Alan R. Liss.

ILAR (Institute of Laboratory Animal Resources). 1970. Gnotobiotes: Standards and Guidelines for the Breeding, Care, and Management of Laboratory Animals. A report of the ILAR Subcommittee on Standards for Gnotobiotes, Committee on Standards. Washington, D.C.: National Academy of Sciences. 52 pp.

ILAR (Institute of Laboratory Animal Resources). 1974. A Guide to Infectious Diseases of Guinea Pigs, Gerbils, Hamsters, and Rabbits. A report of the ILAR Committee on Laboratory Animal Diseases. ILAR News 17:ID1-ID16.

ILAR (Institute of Laboratory Animal Resources). 1976. Guide for the Care and Use of the Nude (Thymus-Deficient) Mouse in Biomedical Research. A report of the ILAR Committee on Care and Use of the "Nude" Mouse. ILAR News 19:M1-M20.

Ingalls, A. M., M. M. Dickie, and G. D. Snell. 1950. Obese, a new mutation in the house mouse. J. Hered. 41:317–318.

Ishida, Y., G. Ueda, K. Noguchi, R. Nagasawa, S. Hirose, H. Sato, and T. Shirai. 1987. Unique cell surface phenotypes of proliferating lymphocytes in mice homozygous for *lpr* and *gld* mutations, defined by monoclonal antibodies to MRL/Mp-*lpr/lpr* T cells. Cell. Immunol. 105:136–146.

Ishigatsubo, Y., D. M. Klinman, and A. D. Steinberg. 1987. Transfer of anti-DNA-producing B cells from NZB to unmanipulated *xid* recipients: Effects of age, sex, and environment. Clin. Immunol. Immunopathol. 45:244–251.

Itaya, T., and Z. Ovary. 1979. Suppression of IgE antibody production in SJL mice. IV. Interaction of primed and unprimed T cells. J. Exp. Med. 150:507–516.

Izui, S., and K. Masuda. 1984. Resistance to tolerance induction is not prerequisite to development of murine SLE. J. Immunol. 133:3010–3014.

Izui, S., P. J. McConahey, A. N. Theofilopoulos, and F. J. Dixon. 1979. Association of circulating retroviral gp70-anti-gp70 immune complexes with murine systemic lupus erythematosus. J. Exp. Med. 149:1099–1116.

Izui, S., K. Masuda, H. Yoshida, V. E. Kelley, J. B. Roths, and I. Hara. 1984. The *lpr* (lymphoproliferation) gene and retroviral gp70 gene in murine lupus. Pp. 45–61 in Recent Advances in Systemic Lupus Erythematosus, P. H. Lambert, L. Perrin, and S. Izui, eds. London: Academic Press.

Jackson, R., N. Rassi, T. Crump, B. Haynes, and G. S. Eisenbarth. 1981. The BB diabetic rat. Profound T-cell lymphocytopenia. Diabetes 30:887–889.

Jackson, R. A., J. B. Buse, R. Rifai, D. Pelletier, E. L. Milford, C. B. Carpenter, G. S. Eisenbarth, and R. M. Williams. 1984. Two genes required for diabetes in BB rats. Evidence from cyclical intercrosses and backcrosses. J. Exp. Med. 159:1629–1636.

Jacob, C. O., P. H. van der Meide, and H. O. McDevitt. 1987. In vivo treatment of (NZB × NZW)F1 lupus-like nephritis with monoclonal antibody to γ interferon. J. Exp. Med. 166:798–803.

Jahrling, P. B., S. Smith, R. A. Hesse, and J. B. Rhoderick. 1982. Pathogenesis of lassa virus infection in guinea pigs. Infect. Immun. 37:771–778.

Jakab, G. J. 1981. Interactions between Sendai virus and bacterial pathogens in the murine lung: A review. Lab. Anim. Sci. 31:170–177.

Jenkinson, E. J., W. van Ewijk, and J. J. T. Owen. 1981. Major histocompatibility complex

antigen expression on the epithelium of the developing thymus in normal and nude mice. J. Exp. Med. 153:280–292.
Jerrells, T. R. 1983. Association of an inflammatory I region-associated antigen-positive macrophage influx and genetic resistance of inbred mice to *Rickettsia tsutsugamushi*. Infect. Immun. 42:549–557.
Jerrells, T. R., and J. V. Osterman. 1982. Role of macrophages in innate and acquired host resistance to experimental scrub typhus infection of inbred mice. Infect. Immun. 37:1066–1073.
Johnson, A. G. 1985. Regulation of the immune system by nucleic acids and polynucleotides. Pp. 107–119 in Biological Response Modifiers. New Approaches to Disease Intervention, P. F. Torrence, ed. Orlando, Fla.: Academic Press.
Johnson, B. C., A. Gajjar, C. Kubo, and R. A. Good. 1986. Calories versus protein in onset of renal disease in NZB × NZW mice. Proc. Natl. Acad. Sci. USA 83:5659–5662.
Johnson, D. A., and H. Meier. 1981. Immune responsiveness of HRS/J mice to syngeneic lymphoma cells. J. Immunol. 127:461–464.
Jones, M. G., and G. Harris. 1985. Prolongation of life in female NZB/NZW (F_1) hybrid mice by cyclosporine A. Clin. Exp. Immunol. 59:1–9.
Jose, D. G., and R. A. Good. 1973a. Quantitative effects of nutritional protein and calorie deficiency upon immune responses to tumors in mice. Cancer Res. 33:807–812.
Jose, D. G., and R. A. Good. 1973b. Quantitative effects of nutritional essential amino acid deficiency upon immune responses to tumors in mice. J. Exp. Med. 137:1–9.
Jyonouchi, H., and P. W. Kincade. 1984. Precocious and enhanced functional maturation of B lineage cells in New Zealand black mice during embryonic development. J. Exp. Med. 159:1277–1282.
Jyonouchi, H., P. W. Kincade, R. A. Good, and G. Fernandes. 1981. Reciprocal transfer of abnormalities in clonable B lymphocytes and myeloid progenitors between NZB and DBA/2 mice. J. Immunol. 127:1232–1235.
Kahn, C. R., D. M. Neville, Jr., and J. Roth. 1973. Insulin-receptor interaction in the obese-hyperglycemic mouse. A model of insulin resistance. J. Biol. Chem. 248:244–250.
Kaiserlian, D., D. Delacroix, and J. F. Bach. 1985. The wasted mutant mouse. I. An animal model of secretory IgA deficiency with normal serum IgA. J. Immunol. 135:1126–1131.
Kaminsky, S. G., I. Nakamura, and G. Cudkowicz. 1983. Selective defect of natural killer cell activity against lymphomas in SJL mice: Low responsiveness to interferon inducers. J. Immunol. 130:1980–1984.
Kaminsky, S. G., I. Nakamura, and G. Cudkowicz. 1985. Genetic control of the natural killer cell activity in SJL and other strains of mice. J. Immunol. 135:665–671.
Kanazawa, Y., K. Komeda, S. Sato, S. Mori, K. Akanuma, and F. Takaku. 1984. Non-obese-diabetic mice: Immune mechanisms of pancreatic β-cell destruction. Diabetologia 27:113–115.
Kaplan, S. S., S. S. Boggs, M. A. Nardi, R. E. Basford, and J. M. Holland. 1978. Leukocyte-platelet interactions in a murine model of Chediak-Higashi syndrome. Blood 52:719–725.
Kappler, J. W., T. Wade, J. White, E. Kushnir, M. Blackman, J. Bill, N. Roehm, and P. Marrack. 1987. A T cell receptor Vβ segment that imparts reactivity to a class II major histocompatibility complex product. Cell 49:263–271.
Karagogeos, D., and H. H. Wortis. 1987. Thymus grafts induce B cell development in nude, X-linked immune deficient mice. Eur. J. Immunol. 17:141–144.
Karagogeos, D., N. Rosenberg, and H. H. Wortis. 1986. Early arrest of B cell development in nude, X-linked immune-deficient mice. Eur. J. Immunol. 16:1125–1130.
Karp, D. R., J. P. Atkinson, and D. C. Shreffler. 1982a. Genetic variation in glycosylation of the fourth component of murine complement. Association with hemolytic activity. J. Biol. Chem. 257:7330–7335.

REFERENCES

Karp, D. R., K. L. Parker, D. C. Shreffler, C. Slaughter, and J. D. Capra. 1982b. Amino acid sequence homologies and glycosylation differences between the fourth component of murine complement and sex-limited protein. Proc. Natl. Acad. Sci. USA 79:6347–6349.

Kastern, W., T. Dyrberg, J. Schøller, and I. Kryspin-Sørensen. 1984. Restriction fragment polymorphisms in the major histocompatibility complex of diabetic BB rats. Diabetes 33:807–809.

Kataoka, S., J. Satoh, H. Fujiya, T. Toyota, R. Suzuki, K. Itoh, and K. Kumagai. 1983. Immunologic aspects of the nonobese diabetic (NOD) mouse. Abnormalities of cellular immunity. Diabetes 32:247–253.

Katz, I. R., R. Asofsky, and G. J. Thorbecke. 1980. Suppression of spontaneous reticulum cell sarcoma development and of syngeneic stimulator cells by anti-μ treatment of SJL/J mice. J. Immunol. 125:1355–1359.

Katz, I. R., J. Chapman-Alexander, E. B. Jacobson, S. P. Lerman, and G. J. Thorbecke. 1981. Growth of SJL/J-derived transplantable reticulum cell sarcoma as related to its ability to induce T-cell proliferation in the host. III. Studies on thymectomized and congenitally athymic SJL mice. Cell. Immunol. 65:84–92.

Katz, M. 1978. Nutrition and the immune response. Pp. 35–36 in Inadvertent Modification of the Immune Response: The Effects of Foods, Drugs, and Environmental Contaminants, I. M. Asher, ed. Proceedings of the Fourth FDA Science Symposium held August 28–30, 1978, in Annapolis, Md. DHHS Pub. No. (FDA) 80-1074. Washington, D.C.: U.S. Department of Health and Human Services.

Kay, M. M. B. 1978. Long term subclinical effects of parainfluenza (SENDAI) infection on immune cells of aging mice. Proc. Soc. Exp. Biol. Med. 158:326–331.

Kazura, J. W., C. Gandola, H. R. Rodman, and A. A. F. Mahmoud. 1979. Deficient production of the lymphokine eosinophil stimulation promoter in chemically induced and mutation diabetes mellitus in mice. J. Immunol. 123:2114–2117.

Kees, U., and R. V. Blanden. 1976. A single genetic element in *H-2K* affects mouse T-cell antiviral function in poxvirus infection. J. Exp. Med. 143:450–455.

Kelley, V. E., and J. B. Roths. 1985. Interaction of mutant *lpr* gene with background strain influences renal disease. Clin. Immunol. Immunopathol. 37:220–229.

Kelley, V. E., and A. Winkelstein. 1980. Age- and sex-related glomerulonephritis in New Zealand white mice. Clin. Immunol. Pathol. 16:142–150.

Kelly, E. M. 1957. Research news: New mutants. Mouse News Lett. 16:36.

Kenny, A. D., W. Toepel, and I. Schour. 1958. Calcium and phosphorus metabolism in the *ia* rat. J. Dent. Res. 37:432–443.

Kim, M.-G., W. Schuler, M. J. Bosma, and K. B. Marcu. 1988. Abnormal recombination of *Igh* D and J gene segments in transformed pre-B cells of *scid* mice. J. Immunol. 141:1341–1347.

Kincade, P. W. 1977. Defective colony formation by B lymphocytes from CBA/N and C3H/HeJ mice. J. Exp. Med. 145:249–263.

Kincade, P. W., G. Lee, T. Watanabe, L. Sun, and M. P. Scheid. 1981. Antigens displayed on murine B lymphocyte precursors. J. Immunol. 127:2262–2268.

Kirchner, H., H. M. Hirt, D. L. Rosenstreich, and S. E. Mergenhagen. 1978. Resistance of C3H/HeJ mice to lethal challenge with herpes simplex virus. Proc. Soc. Exp. Biol. Med. 157:29–32.

Kirkland, T. N., and J. Fierer. 1983. Inbred mouse strains differ in resistance to lethal *Coccidioides immitis* infection. Infect. Immun. 40:912–916.

Kitamura, Y., S. Go, and K. Hatanaka. 1978. Decrease of mast cells in W/W^v mice and their increase by bone marrow transplantation. Blood 52:447–452.

Kitamura, Y., M. Yokoyama, H. Matsuda, T. Ohno, and K. J. Mori. 1981. Spleen colony-forming cell as common precursor for tissue mast cells and granulocytes. Nature (London) 291:159–160.

Klassen, L. W., R. S. Krakauer, and A. D. Steinberg. 1977. Selective loss of suppressor cell function in New Zealand mice induced by NTA. J. Immunol 119:830–837.

Klein, J. O., G. M. Green, J. G. Tilles, E. H. Kass, and M. Finland. 1969. Effect of intranasal reovirus infection on antibacterial activity of mouse lung. J. Infect. Dis. 119:43–50.

Klöting, I., O. Stark, and H.-J. Hahn. 1984. Animal model of the insulin-dependent *Diabetes mellitus* in BB rats: Their RT1^u homogeneity and prolonged survival of allogeneic skin grafts. Folia Biol. (Prague) 30:24–32.

Knight, J. G., and D. D. Adams. 1978. Three genes for lupus nephritis in NZB × NZW mice. J. Exp. Med. 147:1653–1660.

Knoblich, A., J. Görtz, V. Härle-Grupp, and D. Falke. 1983. Kinetics and genetics of herpes simplex virus-induced antibody formation in mice. Infect. Immun. 39:15–23.

Koenig, S., M. K. Hoffmann, and L. Thomas. 1977. Induction of phenotypic lymphocyte differentiation in LPS unresponsive mice by an LPS-induced serum factor and by lipid A-associated protein. J. Immunol. 118:1910–1911.

Koevary, S., A. Rossini, W. Stoller, W. Chick, and R. M. Williams. 1983. Passive transfer of diabetes in the BB/W rat. Science 220:727–728.

Koller, L. D. 1981. Modulation of the immune response in laboratory animals by lead and cadmium. Pp. 190–198 in Biological Relevance of Immune Suppression as Induced by Genetic, Therapeutic and Environmental Factors, J. H. Dean and M. Padarathsingh, eds. Proceedings of a technology assessment workshop held November 13–14, 1979, in Williamsburg, Va. New York: Van Nostrand Reinhold.

Koller, L. D., and J. A. Brauner. 1977. Decreased B-lymphocyte response after exposure to lead and cadmium. Toxicol. Appl. Pharmacol. 42:621–624.

Koller, L. D., and J. G. Roan. 1980. Response of lymphocytes from lead, cadmium, and methylmercury exposed mice in the mixed lymphocyte culture. J. Environ. Pathol. Toxicol. 4:393–398.

Koller, L. D., J. H. Exon, and J. G. Roan. 1975. Antibody suppression by cadmium. Arch. Environ. Health 30:598–601.

Koller, L. D., J. G. Roan, and N. I. Kerkvliet. 1979. Mitogen stimulation of lymphocytes in CBA mice exposed to lead and cadmium. Environ. Res. 19:177–188.

Komschlies, K. L., D. L. Greiner, L. Shultz, and I. Goldschneider. 1987. Defective lymphopoiesis in the bone marrow of motheaten (*me/me*) and viable motheaten (*me^v/me^v*) mutant mice. III. Normal mouse bone marrow cells enable *me^v/me^v* prothymocytes to generate thymocytes after intravenous transfer. J. Exp. Med. 166:1162–1167.

Kotzin, B. L., and E. Palmer. 1987. The contribution of NZW genes to lupus-like disease in (NZB × NZW)F$_1$ mice. J. Exp. Med. 165:1237–1251.

Kotzin, B. L., V. L. Barr, and E. Palmer. 1985. A large deletion within the T-cell receptor beta-chain gene complex in New Zealand white mice. Science 229:167–171.

Krakauer, R. S., T. A. Waldmann, and W. Strober. 1976. Loss of suppressor T cells in adult NZB/NZW mice. J. Exp. Med. 144:662–673.

Kulstad, R., ed. 1986. AIDS. Papers from *Science*, 1982–1985. Pub. No. 85-23. Washington, D.C.: American Association for the Advancement of Science.

Kumar, R. K. 1983. Animal model of human disease: Hodgkin's disease: SJL/J murine lymphoma. Am. J. Pathol. 110:393–396.

Kyogoku, M. 1977. Studies on SL/Ni mouse: Animal model of polyarteritis nodosa. Pp. 356–366 in Vascular Lesions of Collagen Diseases and Related Conditions, Y. Shiokawa, ed. Tokyo: University of Tokyo Press.

Kyogoku, M. 1980. Pathogenesis of vasculitis in the SL/Ni Mouse. Pp. 281–294 in Systemic

Lupus Erythematosus, M. Fukase, ed. Japan Medical Research Foundation Pub. No. 11. Tokyo: University of Tokyo Press.

Kyogoku, M., M. Kawashima, M. Nose, N. Yaginuma, and K. Nagao. 1981. Pathological studies on the kidney of systemic arteritis. II. Animal model: Comparative studies on the pathogenesis of arteritis and glomerulonephritis of SL/Ni and MRL/1. Jpn. J. Nephrol. 23:913–920.

Lambert, P. H., and F. J. Dixon. 1968. Pathogenesis of glomerulonephritis of NZB/W mice. J. Exp. Med. 127:507–521.

Landi, M. S., J. W. Kreider, C. M. Lang, and L. P. Bullock. 1982. Effects of shipping on the immune function in mice. Am. J. Vet. Res. 43:1654–1657.

Lane, H. C., J. M. Depper, W. C. Greene, G. Whalen, T. A. Waldmann, and A. S. Fauci. 1985. Qualitative analysis of immune function in patients with the acquired immunodeficiency syndrome. Evidence for a selective defect in soluble antigen recognition. N. Engl. J. Med. 313:79–84.

Lane, P. W. 1962. Research news. Mouse News Lett. 26:35.

Lane, P. W. 1968. New allele. Mouse News Lett. 38:24.

Lane, P. W. 1971. Linkage data—Grey-lethal (*gl*). Mouse News Lett. 44:30.

Lane, P. W. 1973. New mutants and linkages. Mouse News Lett. 48:35.

Lane, P. W. 1979. Linkage data. Mouse News Lett. 60:50.

Lane, P. W. 1981. New mutations and linkages. Mouse News Lett. 65:27.

Lane, P. W., and E. D. Murphy. 1972. Susceptibility to spontaneous pneumonitis in an inbred strain of beige and satin mice. Genetics 72:451–460.

Laskin, C. A., G. Haddad, and C. A. Soloninka. 1986. The regulatory role of NZB T lymphocytes in the production of anti-DNA antibodies in vitro. J. Immunol. 137:1867–1873.

Lee, S.-K., and R. T. Woodland. 1985. Selective effect of irradiation on responses of thymus-independent antigen. J. Immunol. 134:761–764.

Leiter, E. H. 1985. Type C retrovirus production by pancreatic beta cells. Association with accelerated pathogenesis in C3H-*db/db* ("diabetes") mice. Am. J. Pathol. 119:22–32.

Leiter, E. H., D. L. Coleman, A. B. Eisenstein, and I. Strack. 1980. A new mutation (db^{3J}) at the diabetes locus in strain 129/J mice. I. Physiological and histological characterization. Diabetologia 19:58–65.

Leiter, E. H., D. L. Coleman, and K. P. Hummel. 1981. The influence of genetic background on the expression of mutations at the diabetes locus in the mouse. III. Effect of *H-2* haplotype and sex. Diabetes 30:1029–1034.

Leiter, E. H., D. L. Coleman, D. K. Ingram, and M. A. Reynolds. 1983. Influence of dietary carbohydrate on the induction of diabetes in C57BL/KsJ-*db/db* diabetes mice. J. Nutr. 113:184–195.

Leiter, E. H., M. Prochazka, D. L. Coleman, D. V. Serreze, and L. D. Shultz. 1986. Genetic factors predisposing to diabetes susceptibility in mice. Pp. 29–36 in The Immunology of Diabetes Mellitus, M. A. Jaworski, G. D. Molnar, R. V. Rajotte, and B. Singh, eds. International Congress Series, no. 717. Proceedings of the International Symposium on the Immunology of Diabetes held June 26–28, 1986, in Edmonton, Alberta, Canada. Amsterdam: Elsevier.

Leiter, E. H., P. H. Le, and D. L. Coleman. 1987a. Susceptibility to *db* gene and streptozotocin-induced diabetes in C57BL mice: Control by gender-associated, MHC-unlinked traits. Immunogenetics 26:6–13.

Leiter, E. H., M. Prochazka, and L. D. Shultz. 1987b. Effect of immunodeficiency on diabetogenesis in genetically diabetic (*db/db*) mice. J. Immunol. 138:3224–3229.

Lerner, A. B., T. Shiohara, R. E. Boissy, K. A. Jacobson, M. L. Lamoreux, and G. E. Moellmann. 1986. A mouse model for vitiligo. J. Invest. Dermatol. 87:299–304.

Levan, G., J. Szpirer, C. Szpirer, and M. C. Yoshida. 1986. Present status of chromosome localization of rat genes. Rat News Lett. 17:3–8.

Levey, R. H., N. Trainin, L. W. Law, P. H. Black, and W. P. Rowe. 1963. Lymphocytic choriomeningitis infection in neonatally thymectomized mice bearing diffusion chambers containing thymus. Science 142:483–485.

Levy, G. A., J. L. Leibowitz, and T. S. Edgington. 1981. Induction of monocyte procoagulant activity by murine hepatitis virus type 3 parallels disease susceptibility in mice. J. Exp. Med. 154:1150–1163.

Lieber, M. R., J. E. Hesse, S. Lewis, G. C. Bosma, N. Rosenberg, K. Mizuuchi, M. J. Bosma, and M. Gellert. 1988. The defect in murine severe combined immune deficiency: Joining of signal sequences but not coding segments in V(D)J recombination. Cell 55:7–16.

Like, A. A., A. A. Rossini, D. L. Guberski, M. C. Appel, and R. M. Williams. 1979. Spontaneous diabetes mellitus: Reversal and prevention in the BB/W rat with antiserum to rat lymphocytes. Science 206:1421–1423.

Like, A. A., R. M. Williams, E. Kislauskis, and A. A. Rossini. 1981. Neonatal thymectomy prevents spontaneous diabetes in the Bio Breeding/Worcester (BB/W) rat. Clin. Res. 29:542A (abstr.).

Like, A. A., V. DlRodi, S. Thomas, D. L. Guberski, and A. A. Rossini. 1984. Prevention of diabetes mellitus in the BB/W rat with Cyclosporin-A. Am J. Pathol. 117:92–97.

Like, A. A., C. A. Biron, E. J. Weringer, K. Byman, E. Sroczynski, and D. L. Guberski. 1986a. Prevention of diabetes in BioBreeding/Worcester rats with monoclonal antibodies that recognize T lymphocytes or natural killer cells. J. Exp. Med. 164:1145–1159.

Like, A. A., D. L. Guberski, and L. Butler. 1986b. Diabetic BioBreeding/Worcester (BB/Wor) rats need not be lymphopenic. J. Immunol. 136:3254–3258.

Lindemann, J. 1964. Inheritance of resistance to influenza virus in mice. Proc. Soc. Exp. Biol. Med. 116:506–509.

Linthicum, D. S., and J. A. Frelinger. 1982. Acute autoimmune encephalomyelitis in mice. II. Susceptibility is controlled by the combination of H-2 and histamine sensitization genes. J. Exp. Med. 156:31–40.

Linthicum, D. S., J. J. Munoz, and A. Blaskett. 1982. Acute experimental autoimmune encephalomyelitis in mice. I. Adjuvant action of *Bordetella pertussis* is due to vasoactive amine sensitization and increased vascular permeability of the central nervous system. Cell. Immunol. 73:299–310.

Lipton, H. L., and M. C. Dal Canto. 1979. Susceptibility of inbred mice to chronic central nervous system infection by Theiler's murine encephalomyelitis virus. Infect. Immun. 26:369–374.

Lipton, H. L., and R. Melvold. 1984. Genetic analysis of susceptibility to Theiler's virus-induced demyelinating disease in mice. J. Immunol. 132:1821–1825.

Lissner, C. R., R. N. Swanson, and A. D. O'Brien. 1983. Genetic control of the innate resistance of mice to *Salmonella typhimurium*: Expression of the *Ity* gene in peritoneal and splenic macrophages isolated in vitro. J. Immunol. 131:3006–3013.

Little, C. C., and A. M. Cloudman. 1937. The occurrence of a dominant spotting mutation in the house mouse. Proc. Natl. Acad. Sci. USA 23:535–537.

Loose, L. D., J. B. Silkworth, and D. W. Simpson. 1978a. Influence of cadmium on the phagocytic and microbicidal activity of murine peritoneal macrophages, pulmonary alveolar macrophages, and polymorphonuclear neutrophils. Infect. Immun. 22:378–381.

Loose, L. D., J. B. Silkworth, and D. Warrington. 1978b. Cadmium-induced phagocyte cytotoxicity. Bull. Environ. Contam. Toxicol. 20:582–588.

Lopez, C. 1975. Genetics of natural resistance to herpesvirus infections in mice. Nature (London) 258:152–153.

Loridon-Rosa, B., H. Haddada, and C. De Vaux St Cyr. 1988. The nude hamster: A potential new animal model for experimental cancer studies [letter to the editor]. Cancer J. 2:112.

Lotzova, E., and J. U. Gutterman. 1979. Effect of glucan on natural killer (NK) cells: Further comparison between NK cells and bone marrow effector cell activities. J. Immunol. 123:607–611.

Lozzio, B. B. 1972. Hematopoiesis in congenitally asplenic mice. Am. J. Physiol. 222:290–295.

Lozzio, B. B. 1976. The lasat mouse: A new model for transplantation of human tissues. Biomedicine 24:144–147.

Lukić, M. L., H. H. Wortis, and S. Leskowitz. 1975. A gene locus affecting tolerance to BGG in mice. Cell. Immunol. 15:457–463.

Luster, M. I., R. E. Faith, J. A. McLachlan, and G. C. Clark. 1979. Effect of in utero exposure to diethylstilbestrol on the immune response in mice. Toxicol. Appl. Pharmacol. 47:279–285.

Luster, M. I., J. A. Blank, and J. H. Dean. 1987. Molecular and cellular basis of chemically induced immunotoxicity. Annu. Rev. Pharmacol. Toxicol. 27:23–49.

Lutzner, M. A., and C. T. Hansen. 1976. Motheaten: An immunodeficient mouse with markedly less ability to survive than the nude mouse in a germfree environment. J. Immunol. 116:1496–1497.

Lutzner, M. A., C. T. Lowrie, and H. W. Jordan. 1967. Giant granules in leukocytes of the beige mouse. J. Hered. 58:299–300.

Lyon, M. F. In press. Rules of nomenclature of inbred strains. In Genetic Variants and Strains of the Laboratory Mouse, 2d ed., M. F. Lyon and A. G. Searle, eds. Oxford: Oxford University Press.

Mahoney, K. H., S. S. Morse, and P. S. Morahan. 1980. Macrophage functions in beige (Chédiak-Higashi syndrome) mice. Cancer Res. 40:3934–3939.

Makino, S., K. Kunimoto, Y. Muraoka, Y. Mizushima, K. Katagiri, and Y. Tochino. 1980. Breeding of a non-obese, diabetic strain of mice. Exp. Anim. 29:1–13.

Makino, S., K. Kunimoto, Y. Muraoka, and K. Katagiri. 1981. Effect of castration on the appearance of diabetes in NOD mouse. Exp. Anim. 30:137–140.

Makino, S., Y. Hayashi, Y. Muraoka, and Y. Tochino. 1985. Establishment of the nonobese-diabetic (NOD) mouse. Pp. 25–32 in Current Topics in Clinical and Experimental Aspects of Diabetes Mellitus, N. Sakamoto, H. K. Min, and S. Baba, eds. Proceedings of the Second Japan-Korea Symposium on Diabetes Mellitus held September 20–21, 1983, in Nagoya, Japan. Amsterdam: Elsevier.

Makino, S., M. Harada, Y. Kishimoto, and Y. Hayashi. 1986. Absence of insulitis and overt diabetes in athymic nude mice with NOD genetic background. Exp. Anim. 35:495–498.

Malynn, B. A., T. K. Blackwell, G. M. Fulop, G. A. Rathbun, A. J. W. Furley, P. Ferrier, L. B. Heinke, R. A. Phillips, G. D. Yancopoulos, and F. W. Alt. 1988. The *scid* defect affects the final step of the immunoglobulin VDJ recombinase mechanism. Cell 54:453–460.

Mandel, M. A., and A. A. F. Mahmoud. 1978. Impairment of cell-mediated immunity in mutation diabetic mice (*db/db*). J. Immunol. 120:1375–1377.

Mann, S. J. 1971. Hair loss and cyst formation in hairless and rhino mutant mice. Anat. Rec. 170:485–499.

Manny, N., S. K. Datta, and R. S. Schwartz. 1979. Synthesis of IgM by cells of NZB and SWR mice and their crosses. J. Immunol. 122:1220–1227.

Marks, C. R., M. F. Seifert, and S. C. Marks III. 1984. Osteoclast populations in congenital osteopetrosis: Additional evidence of heterogeneity. Metab. Bone Dis. Rel. Res. 5:259–264.

Marks, S. C., Jr. 1973. Pathogenesis of osteopetrosis in the *ia* rat: Reduced bone resorption due to reduced osteoclast function. Am. J. Anat. 138:165–189.

Marks, S. C., Jr. 1976a. Tooth eruption and bone resorption: Experimental investigation of the *ia* (osteopetrotic) rat as a model for studying their relationships. J. Oral Pathol. 5:149–163.

Marks, S. C., Jr. 1976b. Osteopetrosis in the *ia* rat cured by spleen cells from a normal littermate. Am. J. Anat. 146:331–338.

Marks, S. C., Jr. 1977. Osteopetrosis in the toothless (*tl*) rat: Presence of osteoclasts but failure to respond to parathyroid extract or to be cured by infusion of spleen or bone marrow cells from normal littermates. Am. J. Anat. 149:289–297.

Marks, S. C., Jr. 1978. Studies of the mechanism of spleen cell cure for osteopetrosis in *ia* rats: Appearance of osteoclasts with ruffled borders. Am. J. Anat. 151:119–130.

Marks, S. C., Jr. 1981. Tooth eruption depends on bone resorption: Experimental evidence from osteopetrotic (*ia*) rats. Metab. Bone Dis. Rel. Res. 3:107–115.

Marks, S. C., Jr. 1982. Morphological evidence for reduced bone resorption in osteopetrotic (*op*) mice. Am. J. Anat. 163:157–167.

Marks, S. C., Jr. 1984. Congenital osteopetrotic mutations as probes of the origin, structure, and function of osteoclasts. Clin. Orthop. Rel. Res. 189:239–263.

Marks, S. C., Jr. 1987. Osteopetrosis—Multiple pathways for the interception of osteoclast function. Appl. Pathol. 5:172–183.

Marks, S. C., Jr., and P. W. Lane. 1976. Osteopetrosis, a new recessive skeletal mutation on chromosome 12 of the mouse. J. Hered. 67:11–18.

Marks, S. C., Jr., and S. N. Popoff. In press. Osteoclast biology in the osteopetrotic (*op*) rat. Am. J. Anat.

Marks, S. C., Jr., and G. B. Schneider. 1978. Evidence for a relationship between lymphoid cells and osteoclasts: Bone resorption restored in *ia* (osteopetrotic) rats by lymphocytes, monocytes and macrophages from a normal littermate. Am. J. Anat. 152:331–341.

Marks, S. C., Jr., and D. G. Walker. 1969. The role of the parafollicular cell of the thyroid gland in the pathogenesis of congenital osteopetrosis in mice. Am. J. Anat. 126:299–314.

Marks, S. C., Jr., and D. G. Walker. 1976. Mammalian osteopetrosis—A model for studying cellular and humoral factors in bone resorption. Pp. 227–301 in The Biochemistry and Physiology of Bone, 2d ed., Vol. 4, G. H. Bourne, ed. New York: Academic Press.

Marks, S. C., Jr., and D. G. Walker. 1981. The hematogenous origin of osteoclasts: Experimental evidence from osteopetrotic (microphthalmic) mice treated with spleen cells from beige mouse donors. Am. J. Anat. 161:1–10.

Marks, S. C., Jr., S. F. Seifert, and J. L. McGuire. 1984. Congenitally osteopetrotic (*op/op*) mice are not cured by transplants of spleen or bone marrow cells from normal littermates. Metab. Bone Dis. Rel. Res. 5:183–186.

Marks, S. C., Jr., M. F. Seifert, and P. W. Lane. 1985. Osteosclerosis, a recessive skeletal mutation on chromosome 19 in the mouse. J. Hered. 76:171–176.

Martinez, D., M. A. Brinton, T. G. Tachovsky, and A. H. Phelps. 1980. Identification of lactate dehydrogenase-elevating virus as the etiological agent of genetically restricted, age-dependent polioencephalomyelitis of mice. Infect. Immun. 27:979–987.

Maruo, K., Y. Ueyama, Y. Kuwahara, K. Hioki, M. Saito, T. Nomura, and N. Tamaoki. 1982. Human tumor xenografts in athymic rats and their age dependence. Br. J. Cancer 45:786–789.

Maruyama, N., K. Ohta, S. Hirose, and T. Shirai. 1980. Genetic studies of autoimmunity in New Zealand mice. Immunol. Lett 2:1–5.

Mathes, L. E., R. G. Olsen, L. C. Hebebrand, E. A. Hoover, and J. P. Schaller. 1978. Abrogation of lymphocyte blastogenesis by a feline leukaemia virus protein. Nature (London) 274:687–689.

Mathes, L. E., R. G. Olsen, L. C. Hebebrand, E. A. Hoover, J. P. Schaller, P. W. Adams, and W. S. Nichols. 1979. Immunosuppressive properties of a virion polypeptide, a 15,000-dalton protein, from feline leukemia virus. Cancer Res. 39:950–955.

Matsumoto, M. 1979. Studies on the antibody producing system of SL/Ni mouse—Studies on SL/Ni mouse, an animal model of PN: III. Ryumachi. 19:297–309 (in Japanese).

Matsuzawa, T., and B. Cinader. 1982. Polymorphism of age-dependent changes in the production of a Thf helper factor. Cell. Immunol. 69:186–191.

May, D., M. F. W. Festing, W. L. Ford, and M. E. Smith. 1977. More on the nude rat. Rat News Lett. 2:14–16.

May, J. E., M. A. Kane, and M. M. Frank. 1972. Host defense against bacterial endotoxemia—Contribution of the early and late components of complement to detoxification. J. Immunol. 109:893–895.

Mayer, T. C. 1970. A comparison of pigment cell development in albino, steel, and dominant-spotting mutant mouse embryos. Dev. Biol. 23:297–309.

McChesney, M. B., and M. B. A. Oldstone. 1987. Viruses perturb lymphocyte functions: Selected principles characterizing virus-induced immunosuppression. Annu. Rev. Immunol. 5:279–304.

McCune, J. M., R. Namikawa, H. Kaneshima, L. D. Shultz, M. Lieberman, and I. L. Weissman. 1988. The SCID-hu mouse: Murine model for the analysis of human hematolymphoid differentiation and function. Science 241:1632–1639.

McGarry, M. P., E. K. Novak, and R. T. Swank. 1986. Progenitor cell defect correctable by bone marrow transplantation in five independent mouse models of platelet storage pool deficiency. Exp. Hematol. 14:261–265.

McIntire, K. R., and L. W. Law. 1967. Abnormal serum immunoglobulins occurring with reticular neoplasms in an inbred strain of mouse. J. Natl. Cancer Inst. 39:1197–1211.

McPherson, C. W. 1963. Reduction of *Pseudomonas aeruginosa* and coliform bacteria in mouse drinking water following treatment with hydrochloric acid or chlorine. Lab. Anim. Care 13:737–744.

Meade, C. J., J. Sheena, and J. Mertin. 1979. Effects of the obese (*ob/ob*) genotype on spleen cell immune function. Int. Arch. Allergy Appl. Immunol. 58:121–127.

Meade, C. J., D. R. Brandon, W. Smith, R. G. Simmonds, S. Harris, and C. Sowter. 1981. The relationship between hyperglycaemia and renal immune complex deposition in mice with inherited diabetes. Clin. Exp. Immunol. 43:109–120.

Medawar, P. B., and E. M. Sparrow. 1956. The effects of adrenocortical hormones, adrenocorticotrophic hormone and pregnancy on skin transplantation immunity in mice. J. Endocrinol. 14:240–256.

Meier, H., D. D. Myers, and R. J. Huebner. 1969. Genetic control by the *hr*-locus of susceptibility and resistance to leukemia. Proc. Natl. Acad. Sci. USA 63:759–766.

Melvold, R. W., D. M. Jokinen, R. L. Knobler, and H. L. Lipton. 1987. Variations in genetic control of susceptibility to Theiler's murine encephalomyelitis virus (TMEV)-induced demyelinating disease. I. Differences between susceptible SJL/J and resistant BALB/c strains map near the T cell β-chain constant gene on chromosome 6. J. Immunol. 138:1429–1433.

Mergenhagen, S. E., and D. H. Pluznik. 1984. Defective responses to lipid A in C3H/HeJ mice: Approaches to an understanding of lipid A-cell interaction. Rev. Infect. Dis. 6:519–523.

Mertin, J. 1976. Effect of polyunsaturated fatty acids on skin allograft survival and primary and secondary cytotoxic response in mice. Transplantation 21:1–4.

Milhaud, G., M.-L. Labat, B. Graf, M. Juster, N. Balmain, R. Moutier, and K. Toyama. 1975. Démonstration cinétique, radiographique et histologique de la guérison de l'ostéopétrose congénitale du Rat. C. R. Acad. Sci. Ser D 280:2485–2488.

Milhaud, G., M.-L. Labat, M. Parant, C. Damais, and L. Chedid. 1977. Immunological defect and its correction in the osteopetrotic mutant rat. Proc. Natl. Acad. Sci. USA 74:339–342.

Miller, B. J., M. C. Appel, J. J. O'Neil, and L. S. Wicker. 1988. Both the Lyt-2$^+$ and L3T4$^+$ T cell subsets are required for the transfer of diabetes in nonobese diabetic mice. J. Immunol. 140:52–58.

Miller, J. F. A. P., and G. F. Mitchell. 1969. Thymus and antigen reactive cells. Transplant. Rev. 1:3–42.

Miller, J. F. A. P., A. H. E. Marshall, and R. G. White. 1962. The immunological significance of the thymus. Adv. Immunol. 2:111–162.

Miller, R. D., and C. E. Calkins. 1988. Active role of T cells in promoting an *in vitro* autoantibody response to self erythrocytes in NZB mice. Immunology 63:625–630.

Miller, S. C., and S. C. Marks, Jr. 1982. Osteoclast kinetics in osteopetrotic (*ia*) rats cured by spleen cell transfers from normal littermates. Calcif. Tissue Int. 34:422–427.

Minato, N., L. Reid, H. Cantor, P. Lengyel, and B. R. Bloom. 1980. Mode of regulation of natural killer cell activity by interferon. J. Exp. Med. 152:124–137.

Minkin, C. 1981. Defective macrophage chemotaxis in osteopetrotic mice. Calcif. Tissue Int. 33:677–678.

Minkin, C., and S. Pokress. 1980. Macrophage function in osteopetrosis: Macrophage chemotaxis in microphthalmic (*mi/mi*) mice. J. Dent. Res. 59(special issue A):333 (abstr. 263).

Minkin, C., G. Trump, and S. Stohlman. 1982. Immune function in congenital osteopetrosis: Defective lymphocyte function in microphthalmic mice. Dev. Comp. Immunol. 6:151–159.

Miyazaki, A., T. Hanafusa, K. Yamada, J. Miyagawa, H. Fujino-Kurihara, H. Nakajima, K. Nonaka, and S. Tarui. 1985. Predominance of T lymphocytes in pancreatic islets and spleen of pre-diabetic non-obese diabetic (NOD) mice: A longitudinal study. Clin. Exp. Immunol. 60:622–630.

Miyazawa, M., M. Nose, M. Kawashima, and M. Kyogoku. 1987. Pathogenesis of arteritis of SL/Ni mice. Possible lytic effects of anti-gp70 antibodies on vascular smooth muscle cells. J. Exp. Med. 166:890–908.

Mjaaland, S., and S. Fossum. 1987. The localization of antigen in lymph node follicles of congenitally athymic nude rats. Scand. J. Immunol. 26:141–147.

Moeller, G. R., L. Terry, and R. Snyderman. 1978. The inflammatory response and resistance to endotoxin in mice. J. Immunol. 120:116–123.

Molomut, N., and M. Padnos. 1965. Inhibition of transplantable and spontaneous murine tumours by the M-P virus. Nature (London) 208:948–950.

Mond, J. J., I. Scher, J. Cossman, S. Kessler, P. K. A. Mongini, C. Hansen, F. D. Finkelman, and W. E. Paul. 1982. Role of the thymus in directing the development of a subset of B lymphocytes. J. Exp. Med. 155:924–936.

Mond, J. J., G. Norton, W. E. Paul, I. Scher, F. D. Finkelman, S. House, M. Schaefer, P. K. A. Mongini, C. Hansen, and C. Bona. 1983. Establishment of an inbred line of mice that express a synergistic immune defect precluding *in vitro* responses to type 1 and type 2 antigens, B cell mitogens, and a number of T cell-derived helper factors. J. Exp. Med. 158:1401–1414.

Monier, J. C., and M. Robert. 1974. Defective T-cell functions in autoimmune Swan mice. Ann. Immunol. (Paris) 125:405–413.

Monier, J. C., and M. Sepetjian. 1975. Spontaneous antinuclear autoimmunisation in Swan and nude mice: Comparative study. Ann. Immunol. (Paris) 126C:63–75.

Monier, J. C., J. Thivolet, A. J. Beyvin, J. C. Czyba, D. Schmitt, and D. Salussola. 1971. Les souris SWAN (Swiss antinucléaires). Modèle animal pour l'étude du lupus érythémateux aigu disséminé et de l'amylose. Étude immunopathologique. Pathol. Eur. 6:357–383.

Monier, J. C., M. Sepetjian, J. C. Czyba, J. P. Ortonne, and J. Thivolet. 1974. Spontaneous autoimmunization in nude mice. Pp. 243–249 in Proceedings of the First International

Workshop on Nude Mice, J. Rygaard and C. O. Povlsen, eds. Proceedings of a workshop held October 11–13, 1973, in Scanticon, Aarhus, Denmark. Stuttgart: Gustav Fischer Verlag.

Moore, H. D. 1919. Complementary and opsonic functions in their relation to immunity. A study of the serum of guinea-pigs naturally deficient in complement. J. Immunol. 4:425–441.

Morelli, R., and L. T. Rosenberg. 1971. Role of complement during experimental *Candida* infection in mice. Infect. Immun. 3:521–523.

Mori, R., T. Tasaki, G. Kimura, and K. Takeya. 1967. Depression of acquired resistance against herpes simplex virus infection in neonatally thymectomized mice. Arch. Gesamte Virusforsch. 21:459–462.

Mori, R., K. Kimoto, and K. Takeya. 1970. The role of the thymus in antibody production and in resistance to Japanese encephalitis virus infection. Arch. Gesamte Virusforsch. 29:32–38.

Morozumi, P. A., J. W. Halpern, and D. A. Stevens. 1981. Susceptibility differences of inbred strains of mice to blastomycosis. Infect. Immun. 32:160–168.

Morrison, D. C., and J. L. Ryan. 1979. Bacterial endotoxins and host immune responses. Adv. Immunol. 28:293–450.

Morrissey, P. J., D. R. Parkinson, R. S. Schwartz, and S. D. Waksal. 1980. Immunologic abnormalities in HRS/J mice. I. Specific deficit in T lymphocyte helper function in a mutant mouse. J. Immunol. 125:1558–1562.

Morse, H. C., III, W. F. Davidson, R. A. Yetter, E. D. Murphy, J. B. Roths, and R. L. Coffman. 1982. Abnormalities induced by the mutant gene *lpr*: Expansion of a unique lymphocyte subset. J. Immunol. 129:2612–2615.

Morse, H. C., III, J. B. Roths, W. F. Davidson, W. Y. Langdon, T. N. Fredrickson, and J. W. Hartley. 1985. Abnormalities induced by the mutant gene, *lpr*. Patterns of disease and expression of murine leukemia viruses in SJL/J mice homozygous and heterozygous for *lpr*. J. Exp. Med. 161:602–616.

Mosbach-Ozmen, L., P. Fonteneau, and F. Loor. 1985a. The C57BL/6-*nu/nu lpr/lpr* mouse. I. Expression of the '*lpr*' phenotype' in the C57BL/6 genetic background. Thymus 7:221–232.

Mosbach-Ozmen, L., P. Fonteneau, and F. Loor. 1985b. The C57BL/6-*nu/nu lpr/lpr* mouse. II. Pedigree and preliminary characteristics. Thymus 7:233–245.

Mosier, D. E., R. J. Gulizia, S. M. Baird, and D. B. Wilson. 1988. Transfer of a functional human immune system to mice with severe combined immunodeficiency. Nature (London) 335:256–259.

Mountz, J. D., W. C. Gause, F. D. Finkelman, and A. D. Steinberg. 1988. Prevention of lymphadenopathy in MRL-*lpr/lpr* mice by blocking peripheral lymph node homing with Mel-14 in vivo. J. Immunol. 140:2943–2949.

Moutier, R., H. Lamendin, and S. Berenholc. 1973. Ostéopétrose par mutation spontanée chez le rat. Exp. Anim. 6:87–101.

Moutier, R., K. Toyama, and M. F. Charrier. 1974. Genetic study of osteopetrosis in the Norway rat. J. Hered. 65:373–375.

Moutier, R., K. Toyama, W. R. Cotton, and J. F. Gaines. 1976. Three recessive genes for congenital osteopetrosis in the Norway rat. J. Hered. 67:189–190.

Müller-Eberhard, H. J. 1975. Complement. Am. Rev. Biochem. 44:697–724.

Muraoka, S., and R. G. Miller. 1988. The autoimmune mouse MRL/Mp-*lpr/lpr* contains cells with spontaneous cytotoxic activity against target cells bearing self-determinants. Cell. Immunol. 113:20–32.

Murphy, E. D. 1963. SJL/J, a new inbred strain of mouse with a high, early incidence of reticulum-cell neoplasms. Proc. Am. Assoc. Cancer Res. 4:46 (abstr.).

Murphy, E. D. 1969. Transplantation behavior of Hodgkin's-like reticulum cell neoplasms of strain SJL/J mice and results of tumor reinoculation. J. Natl. Cancer Inst. 42:797–814.

Murphy, E. D. 1979. Reticulum cell sarcomas: Mouse. Pp. 169–173 in Inbred and Genetically Defined Strains of Laboratory Animals. Part 1. Mouse and Rat, P. L. Altman and D. D. Katz, eds. Bethesda, Md.: Federation of American Societies for Experimental Biology.

Murphy, E. D. 1981. Lymphoproliferation (*lpr*) and other single-locus models for murine lupus. Pp. 143–173 in Immunologic Defects in Laboratory Animals, Vol. 2, M. E. Gershwin and B. Merchant, eds. New York: Plenum.

Murphy, E. D., and J. B. Roths. 1977. A single gene model for massive lymphoproliferation with immune complex disease in the new mouse strain MRL. Pp. 69–72 in Topics in Hematology, S. Seno, F. Takaku, and S. Irino, eds. Proceedings of the 16th International Congress of Hematology, held September 5–11, 1976, in Kyoto, Japan. Amsterdam: Excerpta Medica.

Murphy, E. D., and J. B. Roths. 1978a. Autoimmunity and lymphoproliferation: Induction by mutant gene *lpr*, and acceleration by a male-associated factor in strain BXSB mice. Pp. 207–221 in Genetic Control of Autoimmune Disease, N. R. Rose, P. E. Bigazzi, and N. L. Warner, eds. Developments in Immunology, Vol. 1. Proceedings of a workshop held July 10–12, 1978, in Bloomfield Hills, Mich. New York: Elsevier/North-Holland.

Murphy, E. D., and J. B. Roths. 1978b. Purkinje cell degeneration, a late effect of beige mutations in mice. Pp. 108–109 in 49th Annual Report: 1977–1978. Bar Harbor, Maine: The Jackson Laboratory.

Murphy, E. D., and J. B. Roths. 1978c. New inbred strains with early onset of autoimmunity and lymphoproliferation controlled by defined genetic factors. Special Bulletin. Bar Harbor, Maine: The Jackson Laboratory.

Murphy, E. D., and J. B. Roths. 1979. A Y chromosome associated factor in strain BXSB producing accelerated autoimmunity and lymphoproliferation. Arthritis Rheum. 22:1188–1194.

Murphy, E. D., D. E. Harrison, and J. B. Roths. 1973. Giant granules of beige mice. A quantitative marker for granulocytes in bone marrow transplantation. Transplantation 15:526–530.

Murphy, E. D., J. B. Roths, and E. M. Eicher. 1982. Generalized lymphoproliferative disease (*gld*). Mouse News Lett. 67:20–21.

Murphy, W. J., V. Kumar, and M. Bennett. 1987. Rejection of bone marrow allografts by mice with severe combined immune deficiency (SCID). Evidence that natural killer cells can mediate the specificity of marrow graft rejection. J. Exp. Med. 165:1212–1217.

Murray, F. T., D. F. Cameron, and J. M. Orth. 1983. Gonadal dysfunction in the spontaneously diabetic BB rat. Metabolism 32(suppl. 1):141–147.

Nacy, C. A., and M. S. Meltzer. 1982. Macrophages in resistance to rickettsial infection: Strains of mice susceptible to the lethal effects of *Rickettsia akari* show defective macrophage rickettsicidal activity in vitro. Infect. Immun. 36:1096–1101.

Nadel, E. M., and V. H. Haas. 1956. Effect of the virus of lymphocytic choriomeningitis on the course of leukemia in guinea pigs and mice. J. Natl. Cancer Inst. 17:221–231.

Nakajima, P. B., S. K. Datta, R. S. Schwartz, and B. T. Huber. 1979. Localization of spontaneously hyperactive B cells of NZB mice to a specific B cell subset. Proc. Natl. Acad. Sci. USA 76:4613–4616.

Nakano, K., and B. Cinader. 1980. Accelerated age-dependent decline in the T suppressor capacity of SJL mice. Eur. J. Immunol. 10:309–316.

Nakayama, K., M. Nonaka, S. Yokoyama, Y. D. Yeul, S.-N. Pattanakitsakul, and M. Takahashi. 1987. Recombination of two homologous MHC class III genes of the mouse (*C4* and *Slp*) that accounts for the loss of testosterone dependence of sex-limited protein expression. J. Immunol. 138:620–627.

REFERENCES

Naot, Y., S. Merchav, E. Ben-David, and H. Ginsburg. 1979. Mitogenic activity of *Mycoplasma pulmonis*. I. Stimulation of rat B and T lymphocytes. Immunology 36:399–406.

Nilsson, U. R., and H. J. Müller-Eberhard. 1965. Isolation of β_{1F}-globulin from human serum and its characterization as the fifth component of complement. J. Exp. Med. 122:277–298.

Nilsson, U. R., and H. J. Müller-Eberhard. 1967. Deficiency of the fifth component of complement in mice with an inherited complement defect. J. Exp. Med. 125:1–16.

Nisbet, N. W., S. F. Waldron, and M. J. Marshall. 1983. Failure of thymic grafts to stimulate resorption of bone in the Fatty/Orl-*op* rat. Calcif. Tissue Int. 35:122–125.

Nishimoto, H., H. Kikutani, K. Yamamura, and T. Kishimoto. 1987. Prevention of autoimmune insulitis by expression of I-E molecules in NOD mice. Nature (London) 328:432–434.

Nishimura, M., and H. Miyamoto. 1987. Immunopathological influence of the A^y, *db*, *ob*, and *nu* genes placed on the inbred NOD background as murine models for human type I diabetes. J. Immunogenet. 14:127–130.

Nishizuka, Y. 1979. SL/Ni mouse: A strain with high incidence of glomerulonephritis and polyarteritis nodosa. Pp. 578–587 in Immunopathologic Diseases, A. Okabayashi, ed. Tokyo: Bunkodo.

Nishizuka, Y., M. Shisa, O. Taguchi, M. Kyogoku, and M. Matsumoto. 1975. Experimental studies on polyarteritis nodosa I. Nippon Byori Gakkai Kaishi (Trans. Soc. Pathol. Jpn.) 64:108–109 (in Japanese).

Nixon-Fulton, J. L., P. L. Witte, R. E. Tigelaar, P. R. Bergstresser, and V. Kumar. 1987. Lack of dendritic Thy-1$^+$ epidermal cells in mice with severe combined immunodeficiency disease. J. Immunol. 138:2902–2905.

Nordeen, S. K., V. G. Schaefer, M. H. Edgell, C. A. Hutchinson III, L. D. Shultz, and M. Swift. 1984. Evaluations of wasted mouse fibroblasts and SV-40 transformed human fibroblasts as models of ataxia telangiectasia in vitro. Mutat. Res. 140:219–222.

Nose, M., M. Kawashima, K. Yamamoto, T. Sawai, N. Yaginuma, K. Nagao, and M. Kyogoku. 1981. Role of immune complex in the pathogenesis of arteritis in SL/Ni mice: Possible effect as accelerator rather than as initiator. Pp. 109–115 in New Horizons in Rheumatoid Arthritis, Y. Shiokawa, T. Abe, and Y. Yamauchi, eds. Proceedings of the International Congress on Rheumatoid Arthritis held August 24–26, 1980, in Hakone, Japan. Amsterdam: Excerpta Medica.

Novak, E. K., M. P. McGarry, and R. T. Swank. 1985. Correction of symptoms of platelet storage pool deficiency in animal models for Chediak-Higashi syndrome and Hermansky-Pudlak syndrome. Blood 66:1196–1201.

NRC (National Research Council). 1985. Guide for the Care and Use of Laboratory Animals. A report of the Institute of Laboratory Animal Resources Committee on Care and Use of Laboratory Animals. NIH Pub. No. 86-23. Washington, D.C.: U.S. Department of Health and Human Services.

Nyberg, L. M., and S. C. Marks, Jr. 1975. Organ culture of osteopetrotic (*ia*) rat bone: Evidence that the defect is cellular. Am. J. Anat. 144:373–378.

O'Brien, A. D., and E. S. Metcalf. 1982. Control of early *Salmonella typhimurium* growth in innately *Salmonella*-resistant mice does not require functional T lymphocytes. J. Immunol. 129:1349–1351.

O'Brien, A. D., D. L. Rosenstreich, I. Scher, G. H. Campbell, R. P. MacDermott, and S. B. Formal. 1980. Genetic control of susceptibility to *Salmonella typhimurium* in mice: Role of the *Lps* gene. J. Immunol. 124:20–24.

O'Brien, A. D., D. L. Rosenstreich, and I. Scher. 1981. Genetic control of murine resistance to *Salmonella typhimurium* infection. Pp. 37–48 in Immunomodulation by Bacteria and Their Products, H. Friedman, T. W. Klein, and A. Szentivanyi, eds. New York: Plenum.

O'Brien, A. D., D. A. Weinstein, M. Y. Soliman, and D. L. Rosenstreich. 1985. Additional evidence that the *Lps* gene locus regulates natural resistance to *S. typhimurium* in mice. J. Immunol. 134:2820–2823.

Ochs, H. D., S. I. Rosenfeld, E. D. Thomas, E. R. Giblett, C. A. Alper, B. Dupont, J. G. Schaller, B. C. Gilliland, J. A. Hansen, and R. J. Wedgewood. 1977. Linkage between the gene (or genes) controlling synthesis of the fourth component of complement and the major histocompatibility complex. N. Engl. J. Med. 296:470–475.

Ochs, H. D., C. G. Jackson, S. R. Heller, and R. J. Wedgwood. 1978. Defective antibody response to a T-dependent antigen in C4 deficient guinea pigs and its correction by addition of C4. Fed. Proc. 37:1477 (abstr.).

Ochs, H. D., R. J. Wedgwood, S. R. Heller, and P. G. Beatty. 1986. Complement, membrane glycoproteins, and complement receptors: Their role in regulation of the immune response. Clin. Immunol. Immunopathol. 40:94–104.

Ogata, R. T., and D. S. Sepich. 1985. Murine sex-limited protein: Complete cDNA sequence and comparison with murine fourth complement component. J. Immunol. 135:4239–4244.

Ohsugi, Y., and M. E. Gershwin. 1979. Studies of congenitally immunologic mutant New Zealand mice. III. Growth of B lymphocyte clones in congenitally athymic (nude) and hereditarily asplenic (*Dh/+*) NZB mice: A primary B cell defect. J. Immunol. 123:1260–1265.

Okazaki, K., S.-I. Nishikawa, and H. Sakano. 1988. Aberrant immunoglobulin gene rearrangement in *scid* mouse bone marrow cells. J. Immunol. 141:1348–1352.

Oldstone, M. B. A. 1988. Prevention of type I diabetes in nonobese diabetic mice by virus infection. Science 239:500–502.

Oldstone, M. B. A., and F. J. Dixon. 1971. Lactic dehydrogenase virus-induced immune complex type of glomerulonephritis. J. Immunol. 106:1260–1263.

Oldstone, M. B. A., and F. J. Dixon. 1972. Inhibition of antibodies to nuclear antigen and to DNA in New Zealand mice infected with lactate dehydrogenase virus. Science 175:784–786.

Oldstone, M. B. A., P. Southern, M. Rodriguez, and P. Lampert. 1984. Virus persists in β cells of islets of Langerhans and is associated with chemical manifestations of diabetes. Science 224:1440–1443.

Olitsky, P. K.; and J. M. Lee. 1953. Biologic properties and variations of reactions of the encephalitogenic agent in nervous tissues. J. Immunol. 71:419–425.

Oliver, C., and E. Essner. 1973. Distribution of anomalous lysosomes in the beige mouse: A homologue of Chediak-Higashi syndrome. J. Histochem. Cytochem. 21:218–228.

Olsen, C. E., S. M. Wahl, L. M. Wahl, A. L. Sandberg, and S. E. Mergenhagen. 1978. Immunological defects in osteopetrotic mice. Pp. 389–398 in Mechanisms of Localized Bone Loss, J. E. Horton, T. M. Tarpley, and W. F. Davis, eds. Proceedings of the first scientific evaluation workshop on localized bone loss, held November 14–15, 1977, in Washington, D.C. Washington, D.C.: Information Retrieval Inc.

O'Neill, H. C., and R. V. Blanden. 1983. Mechanisms determining innate resistance to ectromelia virus infection in C57BL mice. Infect. Immun. 41:1391–1394.

Onodera, T., A. Toniolo, U. R. Ray, A. B. Jenson, R. A. Knazek, and A. L. Notkins. 1981. Virus-induced diabetes mellitus. XX. Polyendocrinopathy and autoimmunity. J. Exp. Med. 153:1457–1473.

Ooi, Y. M., and H. R. Colten. 1979. Genetic defect in secretion of complement C5 in mice. Nature (London) 282:207–208.

Örn, A., E. M. Hakansson, M. Gidlund, U. Ramstedt, I. Axberg, H. Wigzell, and L.-G. Lundin. 1982. Pigment mutations in the mouse which also affect lysosomal functions lead to suppressed natural killer cell activity. Scand. J. Immunol. 15:305–310.

Osborn, J. E. 1986. Cytomegalovirus and other herpesviruses of mice and rats. Pp. 421–450 in Viral and Mycoplasmal Infections of Laboratory Rodents. Effects on Biomedical Research,

P. N. Bhatt, R. O. Jacoby, H. C. Morse III, and A. E. New, eds. Orlando, Fla.: Academic Press.

Otis, A. P., and H. L. Foster. 1983. Management and design of breeding facilities. Pp. 17–35 in The Mouse in Biomedical Research. Vol. III: Normative Biology, Immunology, and Husbandry. New York: Academic Press.

Owens, M. H., and B. Bonavida. 1976. Immune functions characteristic of SJL/J mice and their association with age and spontaneous reticulum cell sarcoma. Cancer Res. 36:1077–1083.

Paavonen, T. 1985. Glucocorticoids enhance the in vitro Ig synthesis of pokeweed mitogen-stimulated human B cells by inhibiting the suppressive effect of $T8^+$ T-cells. Scand. J. Immunol. 21:63–71.

Painter, C. J., M. Monestier, A. Chew, A. Bona-Dimitriu, K. Kasturi, C. Bailey, V. E. Scott, C. L. Sidman, and C. A. Bonfa. 1988. Specificities and V genes encoding monoclonal autoantibodies from viable motheaten mice. J. Exp. Med. 167:1137–1153.

Pakes, S. P., Y.-S. Lu, and P. C. Meunier. 1984. Factors that complicate animal research. Pp. 649–665 in Laboratory Animal Medicine, J. G. Fox, B. J. Cohen, and F. M. Loew, eds. Orlando, Fla.: Academic Press.

Palkowski, M. R., M. L. Nordlund, L. A. Rheins, and J. J. Nordlund. 1987. Langerhans' cells in hair follicles of the depigmenting C57BL/Ler-*vit/vit* mouse. A model for human vitiligo. Arch. Dermatol. 123:1022–1028.

Parker, J. C., M. D. Whiteman, and C. B. Richter. 1978. Susceptibility of inbred and outbred mouse strains to Sendai virus and prevalence of infection in laboratory rodents. Infect. Immun. 19:123–130.

Pasko, K. L., S. B. Salvin, and A. Winkelstein. 1981. Mechanisms in the *in vivo* release of lymphokines. V. Responses in alloxan-treated and genetically diabetic mice. Cell. Immunol. 62:205–219.

Patel, F., and J. O. Minta. 1979. Biosynthesis of a single chain pro-C5 by normal mouse liver mRNA: Analysis of the molecular basis of C5 deficiency in AKR/J mice. J. Immunol. 123:2408–2414.

Pattengale, P. K., and C. R. Taylor. 1983. Experimental models of lymphoproliferative disease: The mouse as a model for human non-Hodgkin's lymphomas and related leukemias. Am. J. Pathol. 113:237–265.

Pearson, D. J., and G. Taylor. 1975. The influence of the nematode *Syphacia obvelata* on adjuvant arthritis in the rat. Immunology 29:391–396.

Peck, R. M., G. J. Eaton, E. B. Peck, and S. Litwin. 1983. Influence of Sendai virus on carcinogenesis in strain A mice. Lab. Anim. Sci. 33:154–156.

Pedersen, E. B., S. Haahr, and S. C. Mogensen. 1983. X-linked resistance of mice to high doses of herpes simplex virus type 2 correlates with early interferon production. Infect. Immun. 42:740–746.

Pedersen, N. C., E. W. Ho, M. L. Brown, and J. K. Yamamoto. 1987. Isolation of a T-lymphocyte virus from domestic cats with an immunodeficiency-like syndrome. Science 235:790–793.

Pelletier, M., S. Montplaisir, M. Dardenne, and J. F. Bach. 1976. Thymic hormone activity and spontaneous autoimmunity in dwarf mice and their littermates. Immunology 30:783–788.

Peltier, A. P. 1982. Complement. Pp. 252–284 in Immunology, 2d ed., J.-F. Bach, ed. New York: John Wiley & Sons.

Percy, D. H., P. E. Hanna, F. Paturzo, and P. N. Bhatt. 1984. Comparison of strain susceptibility to experimental sialodacryoadenitis in rats. Lab. Anim. Sci. 34:255–260.

Phillips, N. E., and P. A. Campbell. 1981. Antigen-primed helper T cell function in CBA/N mice is radiosensitive. J. Immunol. 127:495–500.

Pickel, K., M. A. Müller, and V. ter Meulen. 1981. Analysis of age-dependent resistance to murine coronavirus JHM infection in mice. Infect. Immun. 34:648–654.

Pierpaoli, W., and N. O. Besedovsky. 1975. Role of the thymus in programming of neuroendocrine functions. Clin. Exp. Immunol. 20:323–338.

Pierro, L. J., and H. B. Chase. 1963. Slate—A new coat color mutant in the mouse. J. Hered. 54:47–50.

Pincus, T. 1980. The endogenous murine type C viruses. Pp. 77–130 in Molecular Biology of RNA Tumor Viruses, J. R. Stephenson, ed. New York: Academic Press.

Pisetsky, D. S., C. Klatt, D. Dawson, and J. B. Roths. 1985. The influence of Yaa on anti-DNA responses of B6-lpr mice. Clin. Immunol. Immunopathol. 37:369–376.

Plant, J., and A. A. Glynn. 1979. Locating salmonella resistance gene on mouse chromosome 1. Clin. Exp. Immunol. 37:1–6.

Plant, J. E., J. M. Blackwell, A. D. O'Brien, J. D. Bradley, and A. A. Glynn. 1982. Are the Lsh and Ity disease resistance genes at one locus on mouse chromosome 1? Nature (London) 297:510–511.

Plant, J. E., G. A. Higgs, and C. S. F. Easmon. 1983. Effects of antiinflammatory agents on chronic *Salmonella typhimurium* infection in a mouse model. Infect. Immun. 42:71–75.

Plotz, P. H. 1983. Autoantibodies are anti-idiotype antibodies to antiviral antibodies. Lancet 2:824–826.

Ponzio, N. M., P. H. Brown, and G. J. Thorbecke. 1986. Host-tumor interactions in the SJL lymphoma model. Int. Rev. Immunol. 1:273–301.

Popovic, M., N. Flomenberg, D. J. Volkman, D. Mann, A. S. Fauci, B. Dupont, and R. C. Gallo. 1984. Alteration of T-cell functions by infection with HTLV-I or HTLV-II. Science 226:459–461.

Potter, M., A. D. O'Brien, E. Skamene, P. Gros, A. Forget, P. A. L. Kongshavn, and J. S. Wax. 1983. A BALB/c congenic strain of mice that carries a genetic locus (Ity^r) controlling resistance to intracellular parasites. Infect. Immun. 40:1234–1235.

Press, J. L. 1981. The CBA/N defect defines two classes of T cell-dependent antigens. J. Immunol. 126:1234–1240.

Prickett, J. D., D. R. Robinson, and A. D. Steinberg. 1981. Dietary enrichment with the polyunsaturated fatty acid eicosapentaenoic acid prevents proteinuria and prolongs survival in NZB × NZW F1 mice. J. Clin. Invest. 68:556–559.

Prochazka, M., E. H. Leiter, D. V. Serreze, and D. L. Coleman. 1987. Three recessive loci required for insulin-dependent diabetes in nonobese diabetic mice. Science 237:286–289.

Prpic, V., J. E. Weiel, S. D. Somers, J. DiGuiseppi, S. L. Gonias, S. V. Pizzo, T. A. Hamilton, B. Herman, and D. O. Adams. 1987. Effects of bacterial lipopolysaccharide on the hydrolysis of phosphatidylinositol-4,5-bisphosphate in murine peritoneal macrophages. J. Immunol. 139:526–533.

Prud'homme, G. J., A. Fuks, E. Colle, T. A. Seemayer, and R. D. Guttmann. 1984. Immune dysfunction in diabetes-prone BB rats. Interleukin 2 production and other mitogen-induced responses are suppressed by activated macrophages. J. Exp. Med. 159:463–478.

Prud'homme, G. J., P. H. Lapchak, N. A. Parfrey, E. Colle, and R. D. Guttmann. 1988. Autoimmunity-prone BB rats lack functional cytotoxic T cells. Cell. Immunol. 114:198–208.

Prueitt, J. L., E. Y. Chi, and D. Lagunoff. 1978. Pulmonary surface-active materials in the Chediak-Higashi syndrome. J. Lipid Res. 19:410–415.

Quimby, F., and L. Dillingham. 1988. Complement. Pp. 145–175 in Clinical Chemistry of Laboratory Animals, W. Loeb and F. Quimby, eds. Elmsford, N.Y.: Pergamon Press.

Quimby, F. W., and R. S. Schwartz. 1982. Systemic lupus erythematosus in mice and dogs. Pp. 1217–1230 in Clinical Aspects of Immunology, 4th ed., Vol. 2, P. J. Lachmann and D. K. Peters, eds. Oxford: Blackwell Scientific.

Rabin, B. S. 1985. Influences on life-span of NZB/W mice. Immunol. Today 6:287 (letter).
Rager-Zisman, B., P. A. Neighbour, G. Ju, and B. R. Bloom. 1980. The role of H-2 in resistance and susceptibility to measles virus infection. Pp. 313–320 in Genetic Control of Natural Resistance to Infection and Malignancy, E. Skamene, P. A. L. Kougshavn, and M. Landy, eds. New York: Academic Press.
Raveche, E. S., S. H. Tjio, W. Boegel, and A. D. Steinberg. 1979. Studies of the effects of sex hormones on autosomal and X-linked genetic control of induced and spontaneous antibody production. Arthritis Rheum. 22:1177–1187.
Rayssiguier, Y., P. Larvor, Y. Augusti, and J. Durlach. 1977. Serum proteins in magnesium-deficient rat. Ann. Biol. Anim. Biochem. Biophys. 17:147–152.
Reddy, S., D. Piccione, H. Takita, and R. B. Bankert. 1987. Human lung tumor growth established in the lung and subcutaneous tissue of mice with severe combined immunodeficiency. Cancer Res. 47:2456–2460.
Reed, N. D., and J. W. Jutila. 1972. Immune responses of congenitally thymusless mice to heterologous erythrocytes. Proc. Soc. Exp. Biol. Med. 139:1234–1237.
Reske-Kunz, A. B., M. P. Scheid, and E. A. Boyse. 1979. Disproportion in T-cell subpopulations in immunodeficient mutant hr/hr mice. J. Exp. Med. 149:228–233.
Rheins, L. A., M. R. Palkowski, and J. J. Nordlund. 1986. Alterations in cutaneous immune reactivity to dinitrofluorobenzene in graying C57BL/vi·vi mice. J. Invest. Dermatol. 86:539–542.
Richter, C. B. 1970. Application of infectious agents to the study of lung cancer: Studies on the etiology and morphogenesis of metaplastic lung lesions in mice. Pp. 365–382 in Morphology of Experimental Respiratory Carcinogenesis, P. Netteshein, M. G. Hanna, Jr., and J. W. Deathekage, eds. Proceedings of a Biology Division, Oak Ridge National Laboratory conference held May 13–16, 1970, in Gatlingburg, Tenn. U.S. Atomic Energy Commission (USAEC) Symposium Series 21. Oak Ridge, Tenn.: USAEC Division of Technical Information.
Riley, V. 1966. Spontaneous mammary tumors: Decrease of incidence in mice infected with the enzyme-elevating virus. Science 153:1657–1658.
Riley, V. 1981. Psychoneuroendocrine influences on immunocompetence and neoplasia. Science 212:1100–1109.
Riley, V., and D. Spackman. 1977. Modifying effects of a benign virus on the malignant process and the role of physiological stress on tumor incidence. Pp. 319–336 in Modulation of Host Immune Resistance in the Prevention or Treatment of Induced Neoplasias, M. A. Chirigos, ed. DHEW Pub. No. (NIH) 77-893. Washington, D.C.: U.S. Department of Health, Education, and Welfare.
Riley, V., D. Spackman, M. A. Fitzmaurice, J. Roberts, J. S. Holcenberg, and W. C. Dolowy. 1974. Therapeutic properties of a new glutaminase-asparaginase preparation and the influence of the lactate dehydrogenase-elevating virus. Cancer Res. 34:429–438.
Riley, V., D. M. Spackman, and G. A. Santisteban. 1978. The LDH virus: An interfering biological contaminant. Science 200:124–126.
Robinson, H. J., H. F. Phares, and O. E. Graessle. 1974. Prostaglandin synthetase inhibitors and infection. Pp. 327–342 in Prostaglandin Synthetase Inhibitors—Their Effect on Physiological Functions and Pathological States, H. J. Robinson and J. R. Vane, eds. New York: Raven Press.
Roder, J., and A. Duwe. 1979. The *beige* mutation in the mouse selectively impairs natural killer cell function. Nature (London) 278:451–453.
Rodriguez, M., J. Leibowitz, and C. S. David. 1986. Susceptibility to Theiler's virus-induced demyelination. Mapping of the gene within the H-2D region. J. Exp. Med. 163:620–631.
Roitt, I., J. Brostoff, and D. Male. 1985. Antibody structure and function. Pp. 5.1–5.10 in Immunology. London: Gower Medical Publishing.

Rolstad, B., S. Fossum, B. Tonessen, and W. L. Ford. 1983. Athymic nude rats resist graft-versus-host disease because they reject allogeneic lymphocytes. Transplant. Proc. 15:1656–1657.

Rose, N. R., A. O. Vladutiu, C. S. David, and D. C. Schreffler. 1973. Autoimmune murine thyroiditis. V. Genetic influence on the disease in BSVS and BRVR mice. Clin. Exp. Immunol. 15:281–287.

Rosenberg, L. T., and D. K. Tachibana. 1962. Activity of mouse complement. J. Immunol. 89:861–867.

Rosenfeld, S. I., L. R. Weitkamp, and F. Ward. 1977. Hereditary C5 deficiency in man: Genetic linkage studies. J. Immunol. 119:604–608.

Rossini, A. A., J. P. Mordes, A. M. Pelletier, and A. A. Like. 1983. Transfusions of whole blood prevent spontaneous diabetes mellitus in the BB/W rat. Science 219:975–977.

Rossini, A. A., J. P. Mordes, and A. A. Like. 1985. Immunology of insulin-dependent diabetes mellitus. Annu. Rev. Immunol. 3:289–320.

Roths, J. B. 1987. Differential expression of murine autoimmunity and lymphoid hyperplasia determined by single genes. Pp. 21–33 in New Horizons in Animal Models for Autoimmune Disease, M. Kyogoku and H. Wigzell, eds. Tokyo: Academic Press.

Roths, J. B., and E. D. Murphy. 1982. The beige mutation retards the progression of autoimmune disease. P. 65 in 53rd Annual Report: 1981–1982. Bar Harbor, Maine: The Jackson Laboratory.

Roths, J. B., E. D. Murphy, and E. M. Eicher. 1984. A new mutation, *gld*, that produces lymphoproliferation and autoimmunity in C3H/HeJ mice. J. Exp. Med. 159:1–20.

Ruco, L. P., M. S. Meltzer, and D. L. Rosenstreich. 1978. Macrophage activation for tumor cytotoxicity: Control of macrophage tumoricidal capacity by the LPS gene. J. Immunol. 121:543–548.

Russell, E. S. 1949. Analysis of pleiotropism at the *W*-locus in the mouse: Relationship between the effects of *W* and W^v substitution on hair pigmentation and on erythrocytes. Genetics 34:708–723.

Russell, E. S. 1970. Abnormalities of erythropoiesis associated with mutant genes in mice. Pp. 649–675 in Regulation of Hematopoiesis, Vol. 1, A. S. Gordon, ed. New York: Appleton-Century-Crofts.

Ryan, J. L., and K. P. W. J. McAdam. 1977. Genetic non-responsiveness of murine fibroblasts to bacterial endotoxin. Nature (London) 269:153–155.

Sabin, A. B. 1952. Nature of inherited resistance to viruses affecting the nervous system. Proc. Natl. Acad. Sci. USA 38:540–546.

Sadoff, D. A., W. E. Giddens, Jr., R. F. DiGiacomo, and A. M. Vogel. 1988. Neoplasms in NIH type II athymic (nude) mice. Lab. Anim. Sci. 38:407–412.

Sainis, K., and S. K. Datta. 1988. $CD4^+$ T cell lines with selective patterns of autoreactivity as well as $CD4^-$ $CD8^-$ T helper cell lines augment the production of idiotypes shared by pathogenic anti-DNA autoantibodies in the NZB × SWR model of lupus nephritis. J. Immunol. 140:2215–2224.

Saito, M., M. Nakagawa, K. Kinoshita, and K. Imaizumi. 1978a. Etiological studies on natural outbreaks of pneumonia in mice. Nippon Juigaku Zasshi (Jpn. J. Vet. Sci.) 40:283–290.

Saito, M., M. Nakagawa, T. Muto, and K. Imaizumi. 1978b. Strain difference of mouse in susceptibility to *Mycoplasma pulmonis* infection. Nippon Juigaku Zasshi (Jpn. J. Vet. Sci.) 40:697–705.

Salzman, L. A., ed. 1986. Animal Models of Retrovirus Infection and Their Relationship to AIDS. Orlando, Fla.: Academic Press. 470 pp.

Sanchez, P., D. Primi, M. Levi-Strauss, and P. A. Cazenave. 1985. The regulatory locus rλ1 affects the level of λ1 light chain synthesis in lipopolysaccharide-activated lymphocytes but not the frequency of λ1-positive B cell precursors. Eur. J. Immunol. 15:66–72.

Sasaki, S., A. Ozawa, and K. Hashimoto, eds. 1981. Recent Advances in Germfree Research. Proceedings of the Seventh International Symposium on Gnotobiology held June 29–July 3, 1981, in Tokyo, Japan. Tokyo: Tokai University Press. 776 pp.

Saxena, R. K., Q. B. Saxena, and W. H. Adler. 1982. Defective T-cell response in beige mutant mice. Nature (London) 295:240–241.

Schaible, R., and J. W. Gowen. 1961. A new dwarf mouse. Genetics 46:896 (abstr.).

Schaller, J. G., B. G. Gilliland, H. D. Ochs, J. P. Leddy, L. C. Y. Agodoa, and S. I. Rosenfeld. 1977. Severe systemic lupus erythematosus with nephritis in a boy with deficiency of the fourth component of complement. Arthritis Rheum. 20:1519–1525.

Scheid, M. P., J. M. Chapman-Alexander, T. Hayama, I. R. Katz, S. P. Lerman, C. Nagler, N. M. Ponzio, and G. J. Thorbecke. 1981. Identification of transplantable reticulum cell sarcomas in SJL/J mice as belonging to the B cell lineage. Pp. 475–482 in B Lymphocytes in the Immune Response: Functional, Developmental, and Interactive Properties, N. Klinman, D. E. Mosier, I. Scher, and E. S. Vitetta, eds. New York: Elsevier/North-Holland.

Scheinberg, M. A., E. S. Cathcart, J. W. Eastcott, M. Skinner, M. Benson, T. Shirahama, and M. Bennett. 1976. The SJL/J mouse: A new model for spontaneous age-associated amyloidosis. I. Morphologic and immunochemical aspects. Lab. Invest. 35:47–54.

Scher, I., M. M. Frantz, and A. D. Steinberg. 1973. The genetics of the immune response to a synthetic double-stranded RNA in a mutant CBA mouse strain. J. Immunol. 110:1396–1401.

Schneider, G. B. 1976. Immunological competence in Snell-Bagg pituitary dwarf mice: Response to contact-sensitizing agent oxazolone. Am. J. Anat. 145:371–393.

Schneider, G. B. 1978. The role of lymphoid cells in bone resorption: Cellular immunological competence in *ia* rats. Am. J. Anat. 153:305–319.

Schneider, G., and S. C. Marks, Jr. 1983. Immunological competence in osteopetrotic (microphthalmic) mice. Exp. Cell Biol. 51:327–336.

Schour, I., S. N. Bhaskar, R. O. Greep, and J. P. Weinmann. 1949. Odontome-like formation in a mutant strain of rats. Am. J. Anat. 85:73–112.

Schuler, W., I. J. Weiler, A. Schuler, R. A. Phillips, N. Rosenberg, T. W. Mak, J. F. Kearney, R. P. Perry, and M. J. Bosma. 1986. Rearrangement of antigen receptor genes is defective in mice with severe combined immune deficiency. Cell 46:963–972.

Schuler, W., A. Schuler, G. G. Lennon, G. C. Bosma, and M. J. Bosma. 1988. Transcription of unrearranged antigen receptor genes in *scid* mice. EMBO J. 7:2019–2024.

Schulman, L. E., J. M. Gumpel, W. A. D'Angelo, R. L. Souhami, M. B. Stevens, A. S. Townes, and A. T. Masi. 1964. Antinuclear factor in inbred strains of mice: The possible role of environmental influence. Arthritis Rheum. 7:753 (abstr.).

Schultz, J. S., S. E. Walker, and C. S. David. 1982. The H-2 haplotype of the PN (Palmerston North) inbred strain. Immunogenetics 16:269–271.

Schwinzer, R., H. J. Hedrich, and K. Wonigeit. 1987. The alloreactive potential of T-like cells from athymic nude rats (*rnu/rnu*). Transplant. Proc. 19:285–287.

Searle, A. G. 1959. Hereditary absence of spleen in the mouse. Nature (London) 184:1419–1420.

Sebesteny, A., and A. C. Hill. 1974. Hepatitis and brain lesions due to mouse hepatitis virus accompanied by wasting in nude mice. Lab. Anim. (London) 8:317–326.

Sedlacek, R. S., H. D. Suit, K. A. Mason, and E. R. Rose. 1980. Development and operation of a stable limited defined flora mouse colony. Pp. 197–201 in Animal Quality and Models in Biomedical Research, A. Spiegel, S. Erichsen, and H. A. Solleveld, eds. Proceedings of the 7th Symposium of the International Council for Laboratory Animal Science (ICLAS) held August 21–23, 1979, in Utrecht, The Netherlands. Stuttgart: Gustav Fischer Verlag.

Seemayer, T. A., E. Colle, G. S. Tannenbaum, L. L. Oligny, R. D. Guttmann, and H. Goldman. 1983. Spontaneous diabetes mellitus syndrome in the rat. III. Pancreatic alterations in aglycosuric and untreated diabetic BB Wistar-derived rats. Metabolism 32(suppl. 1):26–32.

Seifert, M. F., and S. C. Marks, Jr. 1985. Morphological evidence of reduced bone resorption in osteosclerotic (oc) mice. Am. J. Anat. 172:141–153.

Seifert, M. F., and S. C. Marks, Jr. 1987. Congenitally osteosclerotic (oc/oc) mice are resistant to cure by transplantation of bone marrow or spleen cells from normal littermates. Tissue Cell 19:29–37.

Seifert, M. F., S. N. Popoff, and S. C. Marks, Jr. 1988. Skeletal biology in the toothless (osteopetrotic) rat. Am. J. Anat. 183:158–165.

Seldin, M. F., J. P. Reeves, C. L. Scribner, J. B. Roths, W. F. Davidson, H. C. Morse III, and A. D. Steinberg. 1987. Effect of xid on autoimmune C3H-gld/gld mice. Cell. Immunol. 107:249–255.

Serreze, D. V., and E. H. Leiter. 1988. Defective activation of T suppressor cell function in nonobese diabetic mice. J. Immunol. 140:3801–3807.

Serreze, D. V., K. Hamaguchi, and E. H. Leiter. 1988a. Defective immunoregulatory signaling in NOD mice. Diabetes 37(suppl. 1):98A (abstr. 392).

Serreze, D. V., E. H. Leiter, E. L. Kuff, P. Jardieu, and K. Ishizka. 1988b. Molecular mimicry between insulin and retroviral antigen p73. Development of cross-reactive antibodies in sera of NOD and C57BL/KsJ-db/db mice. Diabetes 37:351–358.

Serreze, D. V., E. H. Leiter, S. M. Worthen, and L. D. Shultz. 1988c. NOD marrow stem cells adoptively transfer diabetes to resistant (NOD × NON)F1 mice. Diabetes 37:252–255.

Shapiro, J., J. Jersky, S. Katzav, M. Feldman, and S. Segal. 1981. Anesthetic drugs accelerate the progression of postoperative metastases of mouse tumors. J. Clin. Invest. 68:678–685.

Sharkis, S. J., W. Wiktor-Jedrzejczak, A. Ahmed, G. W. Santos, A. McKee, and K. W. Sell. 1978. Antitheta-sensitive regulatory cell (TSRC) and hematopoiesis: Regulation of differentiation of transplanted stem cells in W/W^v anemic and normal mice. Blood 52:802–817.

Sheena, J., and C. J. Meade. 1978. Mice bearing the ob/ob mutation have impaired immunity. Int. Arch. Allergy Appl. Immun. 57:263–268.

Sherr, D. H., N.-K. Cheung, K. M. Heghinian, B. Benacerraf, and M. E. Dorf. 1979. Immune suppression in vivo with antigen-modified syngeneic cells. II. T cell-mediated nonresponsiveness to fowl γ-globulin. J. Immunol. 122:1899–1904.

Sherr, D. H., M. E. Dorf, M. Gibson, amd C. L. Sidman. 1987. Ly-1 B helper cells in autoimmune "viable motheaten" mice. J. Immunol. 139:1811–1817.

Shevach, E., I. Green, and M. M. Frank. 1976. Linkage of C4 deficiency to the major histocompatibility locus in the guinea pig. J. Immunol. 116:1750 (abstr.).

Shirai, T. K., and R. C. Mellors. 1971. Natural thymocytotoxic autoantibody and reactive antigen in New Zealand black and other mice. Proc. Natl. Acad. Sci. USA 68:1412–1415.

Shirai, T., T. Yoshiki, and R. C. Mellors. 1972. Thymus dependence of cells in peripheral lymphoid tissues and in the circulation sensitive to natural thymocytotoxic autoantibody in NZB mice. J. Immunol. 109:32–37.

Shirai, T., K. Hayakawa, K. Okumura, and T. Tada. 1978. Differential cytotoxic effect of natural thymocytotoxic autoantibody of NZB mice on functional subsets of T cells. J. Immunol. 120:1924–1929.

Shirai, T., S. Hirose, K. Ohta, and N. Maruyama. 1984. H-2 linked genes and autoimmune disease in New Zealand mice. Pp. 75–84 in Immunogenetics. Its Application to Clinical Medicine, T. Sasasuki and T. Tada, eds. Tokyo: Academic Press.

Shirai, T., S. Hirose, I. Sekigawa, T. Okada, and H. Sato. 1987. Genetic and cellular basis of anti-DNA antibody synthesis in systemic lupus erythematosus of New Zealand mice. J. Rheumatol. 14(suppl. 13):11–20.

Shizuru, J. A., C. Taylor-Edwards, B. A. Banks, A. K. Gregory, and C. G. Fathman. 1988. Immunotherapy of the nonobese diabetic mouse: Treatment with an antibody to T-helper lymphocytes. Science 240:659–662.

Shreffler, D. C. 1982. MHC-linked complement components. Pp. 187–219 in Histocompatibility Antigens. Structure and Function. Receptors and Recognition, Series B, Vol. 14, P. Parham and J. Strominger, eds. London: Chapman and Hall.

Shreffler, D. C., and R. D. Owen. 1963. A serologically detected variant in mouse serum: Inheritance and association with the histocompatibility-2 locus. Genetics 48:9–25.

Shultz, L. D., and M. C. Green. 1976. Motheaten, an immunodeficient mutant of the mouse. II. Depressed immune competence and elevated serum immunoglobulins. J. Immunol. 116:936–943.

Shultz, L. D., and C. L. Sidman. 1987. Genetically determined murine models of immunodeficiency. Annu. Rev. Immunol. 5:367–403.

Shultz, L. D., and R. B. Zurier. 1978. "Motheaten": A single gene model for stem cell dysfunction and early onset autoimmunity. Pp. 229–240 in Genetic Control of Autoimmune Disease. Developments in Immunology, Vol. 1, N. R. Rose, P. E. Bigazzi, and N. L. Warner, eds. Proceedings of a workshop held July 10–12, 1978, in Bloomfield Hills, Mich. New York: Elsevier/North-Holland.

Shultz, L. D., H.-J. Heiniger, and E. M. Eicher. 1978. Immunopathology of streaker mice: A remutation to nude in the AKR/J strain. Pp. 211–222 in Animal Models of Comparative and Developmental Aspects of Immunity and Disease, M. E. Gershwin and E. L. Cooper, eds. Proceedings of a symposium held during the annual meeting of the American Society of Zoologists, December 27–30, 1977, in Toronto, Ontario, Canada. Elmsford, N.Y.: Pergamon.

Shultz, L. D., H. G. Bedigian, H.-J. Heiniger, and E. M. Eicher. 1982a. The congenitally athymic streaker mouse. Pp. 33–39 in Proceedings of the Third International Workshop on Nude Mice. Vol. 1: Invited Lectures, Infection, Immunology, N. D. Reed, ed. Proceedings of a workshop held September 6–9, 1979, in Bozeman, Mont. New York: Gustav Fischer Verlag.

Shultz, L. D., H. O. Sweet, M. T. Davisson, and D. R. Coman. 1982b. 'Wasted', a new mutant of the mouse with abnormalities characteristic of ataxia telangiectasia. Nature (London) 297:402–404.

Shultz, L. D., H. Bedigian, G. Carlson, and D. R. Coman. 1983. Effect of congenital athymia on expression of preleukemic cells. Pp. 225–226 in Leukemia Reviews International, Vol. 1, M. A. Rich, ed. New York: Marcel Dekker.

Shultz, L. D., D. R. Coman, C. L. Bailey, W. G. Beamer, and C. L. Sidman. 1984. "Viable motheaten," a new allele at the motheaten locus. I. Pathology. Am. J. Pathol. 116:179–192.

Sidman, C. L., L. D. Shultz, and E. R. Unanue. 1978. The mouse mutant "motheaten." I. Development of lymphocyte populations. J. Immunol. 121:2392–2398.

Sidman, C. L., J. D. Marshall, W. G. Beamer, J. H. Nadeau, and E. R. Unanue. 1986a. Two loci affecting B cell responses to B cell maturation factors. J. Exp. Med. 163:116–128.

Sidman, C. L., L. D. Shultz, R. R. Hardy, K. Hayakawa, and L. A. Herzenberg. 1986b. Production of immunoglobulin isotypes by Ly-1$^+$ B cells in viable motheaten and normal mice. Science 232:1423–1425.

Sidman, R. L., M. C. Green, and S. H. Appel. 1965. Lethargic, *lh*, recessive. P. 34 in Catalog of the Neurological Mutants of the Mouse. Cambridge, Mass.: Harvard University Press.

Siegel, B. V., and J. I. Morton. 1966. Depressed antibody response in the mouse infected with Rauscher leukaemia virus. Immunology 10:559–562.

Siegler, R., and M. A. Rich. 1968. Pathogenesis of reticulum cell sarcoma in mice. J. Natl. Cancer Inst. 41:125–143.

Simmons, M. L., C. B. Richter, R. W. Tennant, and J. Franklin. 1968. Production of specific pathogen-free rats in plastic germfree isolator rooms. Pp. 38–47 in Advances in Germfree Research and Gnotobiology, M. Miyakawa and T. D. Luckey, eds. Proceedings of a symposium on Gnotobiotic Life in Medical and Biological Research held April 6–9, 1967, in Nagoya and Inuyama, Japan. Cleveland: CRC Press.

Singer, P. A., R. J. McEvilly, D. J. Noonan, F. J. Dixon, and A. N. Theofilopoulos. 1986. Clonal diversity and T-cell receptor β-chain variable gene expression in enlarged lymph nodes of MRL-*lpr/lpr* lupus mice. Proc. Natl. Acad. Sci. USA 83:7018–7022.

Skamene, E., P. Gros, A. Forget, P. J. Patel, and M. N. Nesbitt. 1984. Regulation of resistance to leprosy by chromosome 1 locus in the mouse. Immunogenetics 19:117–124.

Skidmore, B. J., J. M. Chiller, W. O. Weigle, R. Riblet, and J. Watson. 1976. Immunologic properties of bacterial lipopolysaccharide (LPS). III. Genetic linkage between the in vitro mitogenic and in vivo adjuvant properties of LPS. J. Exp. Med. 143:143–150.

Slavin, S., S. Strober, Z. Fuks, and H. S. Kaplan. 1977. Induction of specific tissue transplantation tolerance using fractionated total lymphoid irradiation in adult mice: Long-term survival of allogeneic bone marrow and skin grafts. J. Exp. Med. 146:34–48.

Small, J. D. 1984. Rodent and lagomorph health surveillance—Quality assurance. Pp. 709–723 in Laboratory Animal Medicine, J. G. Fox, B. J. Cohen, and F. M. Loew, eds. Orlando, Fla.: Academic Press.

Smith, H. R., T. M. Chused, and A. D. Steinberg. 1983. The effect of the X-linked immune deficiency gene (*xid*) upon the Y chromosome-related disease of BXSB mice. J. Immunol. 131:1257–1262.

Smith, H. R., L. J. Yaffe, D. L. Kastner, and A. D. Steinberg. 1986. Evidence that Lyb-5 is a differentiation antigen in normal and *xid* mice. J. Immunol. 136:1194–1200.

Snell, G. D. 1929. Dwarf, a new Mendelian recessive character of the house mouse. Proc. Natl. Acad. Sci. USA 15:733–734.

Snyderman, R., and G. J. Cianciolo. 1984. Immunosuppressive activity of the retroviral envelope protein P15E and its possible relationship to neoplasia. Immunol. Today 5:240–244.

Soll, A. H., C. R. Kahn, D. M. Neville, Jr., and J. Roth. 1975. Insulin receptor deficiency in genetic and acquired obesity. J. Clin. Invest. 56:769–780.

Sorensen, O., R. Dugre, D. Percy, and S. Dales. 1982. In vivo and in vitro models of demyelinating disease: Endogenous factors influencing demyelinating disease caused by mouse hepatitis virus in rats and mice. Infect. Immun. 37:1248–1260.

Sparrow, S. 1980. The importance of disease in immunodeficient mice and rats. Pp. 25–41 in Immunodeficient Animals for Cancer Research, S. Sparrow, ed. New York: Oxford University Press.

Sprent, J., and J. Bruce. 1984. Physiology of B cells in mice with X-linked immunodeficiency (*xid*). III. Disappearance of *xid* B cells in double bone marrow chimeras. J. Exp. Med. 160:711–723.

Staats, J. 1981. List of inbred strains. Pp. 373–376 in Genetic Variants and Strains of the Laboratory Mouse, M. C. Green, ed. Stuttgart: Gustav Fischer Verlag.

Stanley, N. F., and R. F. Joske. 1975. Animal model of human disease: Active chronic hepatitis. Animal model: Chronic murine hepatitis induced by reovirus type 3. Am. J. Pathol. 80:181–184.

Stanley, P. G. 1965. Non-oncogenic infectious agents associated with experimental tumors. Prog. Exp. Tumor Res. 7:224–258.

Stefansson, K., M. E. Dieperink, D. P. Richman, C. M. Gomez, and L. S. Marton. 1985. Sharing of antigenic determinants between the nicotinic acetylcholine receptor and proteins in *Escherichia coli, Proteus vulgaris* and *Klebsiella pneumoniae*. N. Engl. J. Med. 312:221–225.

Stein, M. 1985. Bereavement, depression, stress, and immunity. Pp. 29–44 in Neural Modulation of Immunity, R. Guillemin, M. Cohn, and T. Melnechuk, eds. Proceedings of a symposium held October 27–28, 1983, under the auspices of the Princess Liliane Cardiology Foundation, Brussels, Belgium. New York: Raven Press.

Steinberg, A. D., J. B. Roths, E. D. Murphy, R. T. Steinberg, and E. S. Raveché. 1980. Effects of thymectomy or androgen administration upon the autoimmune disease of MRL/ Mp-*lpr/lpr* mice. J. Immunol. 125:871–873.

Steinberg, A. D., E. S. Raveché, C. A. Laskin, M. L. Miller, and R. J. Steinberg. 1982. Genetic, environmental and cellular factors in the pathogenesis of systemic lupus erythematosus. Arthritis Rheum. 25:734–743.

Steinberg, A. D., E. S. Reveché, C. A. Laskin, H. R. Smith, T. Santoro, M. L. Miller, and P. H. Plotz. 1984. Systemic lupus erythematosus: Insights from animal models. Ann. Intern. Med. 100:714–727.

Steinberg, A. D., D. M. Klinman, D. L. Kastner, M. F. Seldin, W. C. Gause, C. L. Scribner, J. L. Britten, J. N. Siegel, and J. D. Mountz. 1987. Genetic and molecular genetic studies of murine and human lupus. J. Rheumatol. 14(suppl. 13):166–176.

Steinberg, E. B., T. J. Santoro, T. M. Chused, P. A. Smathers, and A. D. Steinberg. 1983. Studies of congenic MRL-*lpr/lpr.xid* mice. J. Immunol. 131:2789–2795.

Sternberg, E. M., and C. W. Parker. 1988. Pharmacologic aspects of lymphocyte regulation. Pp. 1–54 in The Lymphocyte: Structure and Function, 2d ed., J. J. Marchalonis, ed. New York: Marcel Dekker.

Sternthal, E., A. A. Like, K. Sarantis, and L. E. Braverman. 1981. Lymphocytic thyroiditis and diabetes in the BB/W rat. A new model of autoimmune endocrinopathy. Diabetes 30:1058–1061.

Stevens, J., and J. F. Loutit. 1982. Mast cells in spotted mutant mice (*W Ph mi*). Proc. R. Soc. London Ser. B 215:405–409.

Stiller, C. R., A. Laupacis, P. A. Keown, C. Gardell, J. Dupre, P. Thibert, and W. Wall. 1983. Cyclosporine: Action, pharmacokinetics, and effect in the BB rat model. Metabolism 32(suppl. 1):69–72.

Stohlman, S. A., and J. A. Frelinger. 1978. Resistance to fatal central nervous system disease by mouse hepatitis virus, strain JHM. I. Genetic analysis. Immunogenetics 6:277–281.

Stohlman, S. A., J. A. Frelinger, and L. P. Weiner. 1980. Resistance to fatal central nervous system disease by mouse hepatitis virus, strain JHM. II. Adherent cell-mediated protection. J. Immunol. 124:1733–1739.

Stoye, J. P., S. Fenner, G. E. Greenoak, C. Moran, and J. M. Coffin. 1988. Role of endogenous retroviruses as mutagens: The hairless mutation of mice. Cell 54:383–391

Strang, C. J., H. S. Auerbach, and F. S. Rosen. 1986. C1s-induced vascular permeability in C2-deficient guinea pigs. J. Immunol. 137:631–635.

Strober, S., S. Slavin, M. Gottlieb, I. Zan-Bar, D. P. King, R. T. Hoppe, Z. Fuks, F. C. Grumet, and H. S. Kaplan. 1979. Allograft tolerance after total lymphoid irradiation (TLI). Immunol. Rev. 46:87–112.

Stromberg, K., R. E. Benveniste, L. O. Arthur, H. Rabin, W. E. Giddens, Jr., H. D. Ochs, W. R. Morton, and C.-C. Tsai. 1984. Characterization of exogenous type D retrovirus from a fibroma of a macaque with simian AIDS and fibromatosis. Science 224:289–292.

Sultzer, B. M. 1968. Genetic control of leucocyte responses to endotoxin. Nature (London) 219:1253–1254.

Surh, C. D., M. E. Gershwin, and A. Ahmed. 1987. A peripheral and central T cell antigen recognized by a monoclonal thymocytotoxic autoantibody from New Zealand black mice. J. Immunol. 138:1421–1428.

Suzuki, T., T. Yamada, T. Fujimura, E. Kawamura, M. Shimizu, R. Yamashita, and K. Nomoto. 1987. Diabetogenic effects of lymphoctye transfusion on the NOD or NOD nude mouse. Pp. 112–116 in Immune Deficient Animals in Biomedical Research, J. Rygaard, N. Brünner, N. Græm, and M. Spang-Thomsen, eds. Basel: Karger.

Sweet, H. O. 1984. Wasted linkage. Mouse News Lett. 71:31.

Taguchi, O., A. Kojima, and Y. Nishizuka. 1985. Experimental autoimmune prostatitis after neonatal thymectomy in the mouse. Clin. Exp. Immunol. 60:123–129.

Takaoki, M., and H. Kawaji. 1980. Impaired antibody response against T-dependent antigens in rhino mice. Immunology 40:27–31.

Takatsu, K., and T. Hamaoka. 1982. DBA/2Ha mice as a model of an X-linked immunodeficiency which is defective in the expression of TRF-acceptor site(s) on B lymphocytes. Immunol. Rev. 64:25–55.

Takatsu, K., N. Harada, Y. Hara, Y. Takahama, G. Yamada, K. Dobashi, and T. Hamaoka. 1985. Purification and physiochemical characterization of murine T cell replacing factor (TRF). J. Immunol. 134:382–389.

Takeda, T., M. Hosokawa, S. Takeshita, M. Irino, K. Higuchi, T. Matsushita, Y. Tomita, K. Yasuhira, H. Hamamoto, K. Shimizu, M. Ishii, and T. Yamamuro. 1981. A new murine model of accelerated senescence. Mech. Ageing Dev. 17:183–194.

Talal, N. 1983. Immune response disorders. Pp. 391–399 in The Mouse in Biomedical Research. Vol. 3: Normative Biology, Immunology, and Husbandry, H. L. Foster, J. D. Small, and J. G. Fox, eds. New York: Academic Press.

Tattersall, P., and S. F. Cotmore. 1986. The rodent parvoviruses. Pp. 305–348 in Viral and Mycoplasmal Infections of Laboratory Rodents. Effects on Biomedical Research, P. N. Bhatt, R. O. Jacoby, H. C. Morse III, and A. E. New, eds. Orlando, Fla.: Academic Press.

Taurog, J. D., E. S. Raveché, P. A. Smathers, L. H. Glimcher, D. P. Huston, C. T. Hansen, and A. D. Steinberg. 1981. T cell abnormalities in NZB mice occur independently of autoantibody production. J. Exp. Med. 153:221–234.

Taylor, C. R. 1976. Immuno-histological observations upon the development of reticulum cell sarcoma in the mouse. J. Pathol. 118:201–219.

Teuscher, C. 1985. Experimental allergic orchitis in mice. II. Association of disease susceptibility with the locus controlling *Bordetella pertussis*-induced sensitivity to histamine. Immunogenetics 22:417–425.

Theofilopoulos, A. N. 1986. Molecular genetics of murine lupus. Agents Actions 19:282–294.

Theofilopoulos, A. N., and F. J. Dixon. 1985. Murine models of systemic lupus erythematosus. Adv. Immunol. 37:269–390.

Theofilopoulos, A. N., R. A. Eisenberg, M. Bourdon, J. S. Crowell, Jr., and F. J. Dixon. 1979. Distribution of lymphocytes identified by surface markers in murine strains with systemic lupus erythematosus-like syndromes. J. Exp. Med. 149:516–534.

Thigpen, J. E., and P. W. Ross. 1983. Viral cross contamination of rats maintained in a fabricwalled mass air flow system. Lab. Anim. Sci. 33:446–450.

Thong, Y. H., and A. Ferrante. 1980. Effect of tetracycline on immunological responses in mice. Clin. Exp. Immunol. 39:728–732.

Thunert, A., L. C. Jonas, S. Rehm, and E. Sickel. 1985. Transmission and course of Tyzzer's disease in euthymic and thymus aplastic nude Han:RNU rats. Z. Versuchstierkd. 27:241–248.

Tisdale, W. A. 1963. Potentiating effect of K virus on mouse hepatitis virus (MHV-5) in weanling mice. Proc. Soc. Exp. Biol. Med. 114:774–777.

Tochino, Y. 1987. The NOD mouse as a model of type I diabetes. CRC Crit. Rev. Immunol. 8:49–81.
Tochino, Y., T. Kanaya, and S. Makino. 1983. Microangiopathy in the spontaneously diabetic nonobese mouse (NOD mouse) with insulitis. Pp. 423–432 in Diabetic Microangiopathy, H. Abe and M. Hoshi, eds. Tokyo: University of Tokyo Press.
Tominaga, A., K. Takatsu, and T. Hamaoka. 1980. Antigen-induced T cell-replacing factor (TRF). II. X-linked gene control for the expression of TRF-acceptor site(s) on B lymphocytes and preparation of specific antiserum to that acceptor. J. Immunol. 124:2423–2429.
Tonietti, G., M. B. A. Oldstone, and F. J. Dixon. 1970. The effect of induced chronic viral infections on the immunologic diseases of New Zealand mice. J. Exp. Med. 132:89–109.
Toyota, T., J. Satoh, K. Oya, S. Shintani, and T. Okano. 1986. Streptococcal preparation (OK-432) inhibits development of type I diabetes in NOD mice. Diabetes 35:496–499.
Trexler, P. C. 1976. Gnotobiotic animals: 1. Pp. 135–146 in The UFAW Handbook on the Care and Management of Laboratory Animals, 5th ed., Universities Federation for Animal Welfare (UFAW), ed. Edinburgh: Churchill Livingstone.
Trexler, P. C. 1983. Gnotobiotics. Pp. 1–16 in The Mouse in Biomedical Research. Vol. III: Normative Biology, Immunology, and Husbandry, H. L. Foster, J. D. Small, and J. G. Fox, eds. New York: Academic Press.
Van Bleek, G. M., M. De Hullu, J. Sanders, D. Hoekstra, J. Calafat, and C. J. Lucas. 1988. Recognition of viral antigens by cytotoxic T lymphocytes. Ann. N.Y. Acad. Sci. 532:429–431.
Van Vliet, E., E. J. Jenkinson, R. Kingston, J. J. T. Owen, and W. Van Ewijk. 1985. Stromal cell types in the developing thymus of the normal and nude mouse embryo. Eur. J. Immunol. 15:675–681.
Vassalli, J.-D., A. Granelli-Piperno, C. Griscelli, and E. Reich. 1978. Specific protease deficiency in polymorphonuclear leukocytes of Chediak-Higashi syndrome and beige mice. J. Exp. Med. 147:1285–1290.
Vladutiu, A. O., and N. R. Rose. 1971. Autoimmune murine thyroiditis in relation to histocompatibility (H-2) type. Science 174:1137–1139.
Vogel, S. N., C. T. Hansen, and D. L. Rosenstreich. 1979. Characterization of a congenitally LPS-resistant, athymic mouse strain. J. Immunol. 122:619–622.
Vogel, S. N., A. C. Weinblatt, and D. L. Rosenstreich. 1981. Inherent macrophage defects in mice. Pp. 327–357 in Immunologic Defects in Laboratory Animals, Vol. 1, M. E. Gershwin and B. Merchant, eds. New York: Plenum.
Vogel, S. N., G. S. Madonna, L. M. Wahl, and P. D. Rick. 1984. Stimulation of spleen cells and macrophages of C3H/HeJ mice by a lipid A precursor derived from *Salmonella typhimurium*. Rev. Infect. Dis. 6:535–541.
Vos, J. G., J. M. Berkvens, and B. C. Kruijt. 1980. The athymic nude rat. I. Morphology of lymphoid and endocrine organs. Clin. Immunol. Immunopathol. 15:213–228.
Vukajlovich, S. W., and D. C. Morrison. 1984. Lipid A-dependent lymphocyte proliferation in "endotoxin-nonresponder" mice. Rev. Infect. Dis. 6:528–531.
Wagner, J. E., and P. J. Manning. 1976. The Biology of the Guinea Pig. New York: Academic Press. 317 pp.
Walker, D. G. 1975a. Bone resorption restored in osteopetrotic mice by transplants of normal bone marrow and spleen cells. Science 190:784–785.
Walker, D. G. 1975b. Spleen cells transmit osteopetrosis in mice. Science 190:785–787.
Walker, D. G. 1975c. Control of bone resorption by hematopoietic tissue. The induction and reversal of congenital osteopetrosis in mice through use of bone marrow and splenic transplants. J. Exp. Med. 142:651–663.
Walker, S. E. 1988. Defective primary and secondary IgG responses to thymic-dependent antigens in autoimmune PN mice. J. Immunol. 140:3426–3433.

Walker, S. E., and J. E. Hewett. 1984. Responses to T-cell and B-cell mitogens in autoimmune Palmerston North and NZB/NZW mice. Clin. Immunol. Immunopathol. 30:469–478.

Walker, S. E., R. H. Gray, M. Fulton, R. D. Wigley, and B. Schnitzer. 1978. Palmerston North mice, a new animal model of systemic lupus erythematosus. J. Lab. Clin. Med. 92:932–945.

Walker, S. E., A. B. Kier, E. C. Siegfried, B. G. Harris, and J. S. Schultz. 1986. Accelerated autoimmune disease and lymphoreticular neoplasms in F1 hybrid PN/NZB and NZB/PN mice. Clin. Immunol. Immunopathol. 39:81–92.

Wallace, G. D., and R. M. L. Buller. 1985. Kinetics of ectromelia virus (mousepox) transmission and clinical response in C57BL/6J, BALB/cByJ and AKR/J inbred mice. Lab. Anim. Sci. 35:41–46.

Wallace, G. D., and R. M. L. Buller. 1986. Ectromelia virus (mousepox): Biology, epizootiology, prevention and control. Pp. 539–556 in Viral and Mycoplasmal Infections of Laboratory Rodents. Effects on Biomedical Research, P. N. Bhatt, R. O. Jacoby, H. C. Morse III, and A. E. New, eds. Orlando, Fla.: Academic Press.

Wanebo, H. J., W. M. Gallmeier, E. A. Boyse, and L. J. Old. 1966. Paraproteinemia and reticulum cell sarcoma in an inbred mouse strain. Science 154:901–903.

Ward, J. M., D. P. Houchens, M. J. Collins, D. M. Young, and R. L. Reagan. 1976. Naturally occurring Sendai virus infection of athymic nude mice. Vet. Pathol. 13:36–46.

Warren, M. K., and S. N. Vogel. 1985. Opposing effects of glucocorticoids on interferon-γ-induced murine macrophage Fc receptor and Ia antigen expression. J. Immunol. 134:2462–2469.

Watanabe, N., S. Kojima, and Z. Ovary. 1976. Suppression of IgE antibody production in SJL mice. I. Nonspecific suppressor T cells. J. Exp. Med. 143:833–845.

Watanabe, R., H. Wege, and V. ter Meulen. 1983. Adoptive transfer of EAE-like lesions from rats with coronavirus-induced demyelinating encephalomyelitis. Nature (London) 305:150–153.

Watson, J. 1977. Differentiation of B lymphocytes in C3H/HeJ mice: The induction of Ia antigens by lipopolysaccharide. J. Immunol. 118:1103–1108.

Watson, J., and R. Riblet. 1974. Genetic control of responses to bacterial lipopolysaccharides in mice. I. Evidence for a single gene that influences mitogenic and immunogenic responses to lipopolysaccharides. J. Exp. Med. 140:1147–1161.

Watson, J., and R. Riblet. 1975. Genetic control of responses to bacterial lipopolysaccharides in mice. II. A gene that influences a membrane component involved in the activation of bone marrow-derived lymphocytes by lipopolysaccharides. J. Immunol. 114:1462–1468.

Watson, J., K. Kelly, M. Largen, and B. A. Taylor. 1978a. The genetic mapping of a defective *Lps* response gene in C3H/HeJ mice. J. Immunol. 120:422–424.

Watson, J., M. Largen, and K. P. W. J. McAdam. 1978b. Genetic control of endotoxic responses in mice. J. Exp. Med. 147:39–49.

Watters, J. W. F., J. D. Locker, H. W. Kunz, and T. J. Gill III. 1987. Polymorphism and mapping of the complement gene *C4* in the rat. Immunogenetics 25:204–206.

Webster, L. T. 1933. Inherited and acquired factors in resistance to infection. J. Exp. Med. 57:793–843.

Webster, L. T. 1937. Inheritance of resistance of mice to enteric bacterial and neurotropic virus infections. J. Exp. Med. 65:261–286.

Wege, H., J. Winter, P. Massa, R. Dörries, and V. ter Meulen. 1987. Coronavirus JHM induced demyelinating disease: Specific domains on the E2-protein are associated with neurovirulence. Pp. 307–320 in Coronaviruses, M. M. C. Lai and S. A. Stohlman, eds. Proceedings of the Third International Coronavirus Symposium held September 14–18, 1986, in Asilomar, Calif. New York: Plenum.

Weidanz, W. P., W. Morges, and H. Warren. 1978. Infection-related immunodepression. Pp.

30-34 in Inadvertent Modification of the Immune Response: The Effects of Foods, Drugs, and Environmental Contaminants, I. M. Asher, ed. Proceedings of the Fourth FDA Science Symposium held August 28-30, 1978, in Annapolis, Md. DHHS Pub. No. (FDA) 80-1074. Washington, D.C.: U.S. Department of Health and Human Services.

Weihe, W. H. 1984. The thermoregulation of the nude mouse. Pp. 140-144 in Immune Deficient Animals, B. Sordat, ed. Basel: Karger.

Weiss, M., S. H. Ingbar, S. Winblad, and D. L. Kasper. 1983. Demonstration of a saturable binding site for thyrotropin in *Yersinia enterocolitica*. Science 219:1331-1333.

Welles, W. L., and J. R. Battisto. 1981. The significance of hereditary asplenia for immunologic competence. Pp. 191-212 in Immunologic Defects in Laboratory Animals, Vol. 1, M. E. Gershwin and B. Merchant, eds. New York: Plenum.

Wicker, L. S., B. J. Miller, and Y. Mullen. 1986. Transfer of autoimmune diabetes mellitus with splenocytes from nonobese diabetic (NOD) mice. Diabetes 35:855-860.

Wicker, L. S., B. J. Miller, L. Z. Coker, S. E. McNally, S. Scott, Y. Mullen, and M. C. Appel. 1987. Genetic control of diabetes and insulitis in the nonobese diabetic (NOD) mouse. J. Exp. Med. 165:1639-1654.

Wicker, L. S., B. J. Miller, A. Chai, M. Terada, and Y. Mullen. 1988. Expression of genetically determined diabetes and insulitis in the nonobese diabetic (NOD) mouse at the level of bone marrow-derived cells. Transfer of diabetes and insulitis to nondiabetic (NOD × B10)F_1 mice with bone marrow cells from NOD mice. J. Exp. Med. 167:1801-1810.

Wikel, S. K. 1979. Acquired resistance to ticks. Expression of resistance by C4-deficient guinea pigs. Am. J. Trop. Med. Hyg. 28:586-590.

Wiktor-Jedrzejczak, W., R. R. Skelly, and A. Ahmed. 1981. Hematopoietic stem cell differentiation and its role in osteopetrosis. Immunologic implications. Pp. 51-77 in Immunologic Defects in Laboratory Animals, Vol. 1, M. E. Gershwin and B. Merchant, eds. New York: Plenum.

Wiktor-Jedrzejczak, W., A. Ahmed, C. Szczylik, and R. R. Skelly. 1982. Hematological characterization of congenital osteopetrosis in *op/op* mouse. J. Exp. Med. 156:1516-1527.

Wiktor-Jedrzejczak, W., J. Grzybowski, A. Ahmed, and L. Kaczmarek. 1983. Osteopetrosis associated with premature thymic involution in grey-lethal mice. In vitro studies of thymic microenvironment. Clin. Exp. Immunol. 52:465-471.

Williams, F. P., Jr., and R. T. Habermann. 1962. A study of bacterial contamination in commercially prepared animal feeds and bedding. Proc. Anim. Care Panel 12:11-14.

Woda, B. A., A. A. Like, C. Padden, and M. L. McFadden. 1986. Deficiency of phenotypic cytotoxic-suppressor T lymphocytes in the BB/W rat. J. Immunol. 136:856-859.

Wolfsy, D., and N. Y. Chiang. 1987. Proliferation of Ly-1 B cells in autoimmune NZB and (NZB × NZW)F1 mice. Eur. J. Immunol. 17:809-814.

Woloski, B. M. R. N. J., E. M. Smith, W. J. Meyer III, G. M. Fuller, and J. E. Blalock. 1985. Corticotropin-releasing activity of monokines. Science 230:1035-1036.

Wonigeit, K., H. J. Hedrich, and R. Schwinzer. 1987. Demonstration of alloreactive T cells and allogeneic skin graft rejection in athymic nude rats. Pp. 85-88 in Immune Deficient Animals in Biomedical Research, J. Rygaard, N. Brünner, N. Græm, and M. Spang-Thomsen, eds. Basel: Karger.

Wortis, H. H., S. Nehlsen, and J. J. Owen. 1971. Abnormal development of the thymus in "nude" mice. J. Exp. Med. 134:681-692.

Wortis, H. H., L. Burkly, D. Hughes, S. Roschelle, and G. Waneck. 1982. Lack of mature B cells in nude mice with X-linked immune deficiency. J. Exp. Med. 155:903-913.

Yang, S. Y., R. Jensen, L. Folke, R. A. Good, and N. K. Day. 1974. Complement deficiency in hamsters. Fed. Proc. 33:795 (abstr.).

Yoon, J.-W., T. Onodera, and A. L. Notkins. 1978. Virus-induced diabetes mellitus. XV.

Beta cell damage and insulin-dependent hyperglycemia in mice infected with Coxsackie virus B4. J. Exp. Med. 148:1068–1080.

Yoon, J.-W., P. R. McClintock, T. Onodera, and A. L. Notkins. 1980. Virus-induced diabetes mellitus. XVIII. Inhibition by a non-diabetogenic variant of encephalomyocarditis virus. J. Exp. Med. 152:878–892.

Yoon, J.-W., E. H. Leiter, D. L. Coleman, M. K. Kim, C. Y. Pak, R. G. McArthur, and D. A. K. Roncari. 1988. Genetic control of organ reactive autoantibody production in mice by obesity (*ob*) diabetes (*db*) genes. Diabetes 37:1287–1293.

Yoshiki, T., T. Hayasaka, R. Fukatsu, T. Shirai, T. Itoh, H. Ikeda, and M. Katagiri. 1979. The structural proteins of murine leukemia virus and the pathogenesis of necrotizing arteritis and glomerulonephritis in SL/Ni mice. J. Immunol. 122:1812–1820.

Yui, K., Y. Hashimoto, S. Wadsworth, and M. I. Greene. 1987. Characterization of Lyt-2$^-$, L3T4$^-$ class I-specific cytolytic clones in C3H-*gld/gld* mice. Implications for functions of accessory molecules and programmed development. J. Exp. Med. 166:1026–1040.

Zerwekh, J. E., S. C. Marks, Jr., and L. McGuire. 1987. Elevated serum 1,25 dihydroxyvitamin D in osteopetrotic mutations in three species. Bone Mineral 2:193–199.

Ziefer, A., T. Jacobs, H. J. Hedrich, and H. M. Seitz. 1984. Pneumocystis carinii-infections in immunosuppressed and in thymusdeficient rats. Zentralbl. Bakteriol. Hyg. A 258:387 (abstr.).

Zinkernagel, R. M., B. Adler, and A. Althage. 1977. The question of derepression of *H-2* specificities in virus-infected cells: Failure to detect specific alloreactive T cells after systemic virus infection or alloantigens detectable by alloreactive T cells on virus-infected target cells. Immunogenetics 5:367–378.

Zinkernagel, R. M., T. Leist, H. Hengartner, and A. Althage. 1985. Susceptibility to lymphocytic choriomeningitis virus isolates correlates directly with early and high cytotoxic T cell activity, as well as with foot pad swelling reaction, and all three are regulated by H-2D. J. Exp. Med. 162:2125–2141.

Zivkovic, D., J. E. Speksnijder, H. Kuil, and W. Seinen. 1985. Immunity to *Babesia* in mice. III. The effects of corticosteroids and anti-thymocyte serum on mice immune to *Babesia rodhaini*. Vet. Immunol. Immunopathol. 9:131–142.

Zurier, R. B. 1982. Prostaglandins, immune responses and murine lupus. Arthritis Rheum. 25:804–809.

Appendix

Hematopoietic Cell Surface Antigens Mentioned in the Text

Cluster of Differentiation (CD) Designation	Monoclonal Antibodies Reacting to Species-Specific Cell Surface Antigen					Distribution	Description
	Human	Mouse	Rat	Guinea Pig	Hamster		
CD1	T6 Leu-6 OKT6	Ly-38				Langerhans' cells, thymocytes	Cortical thymocyte differentiation marker
CD2	T11 OKT11 Leu-5b	Ly-37	OX34			All T cells forming E rosettes, NK cells	E-rosette receptor
CD3	T3 OKT3 Leu-4					Mature T cells	Constant portion of Ti receptor
CD4	T4 OKT4 Leu-3a	Ly-4 L3T4	OX36 OX37 OX38 W3/25 OX35			Helper/inducer T lymphocytes	A member of the class II-restricted T-cell population
CD5	T1 OKT1 Leu-1	Ly-1 Lyt-1	R-1-3133 OX19	8BE6		T cells, B-cell subset	
CD6	T12					Mature T cells, Subpopulation of B cells	

Continued

213

Hematopoietic Cell Surface Antigens Mentioned in the Text (*continued*)

Cluster of Differentiation (CD) Designation	Monoclonal Antibodies Reacting to Species-Specific Cell Surface Antigen					Distribution	Description
	Human	Mouse	Rat	Guinea Pig	Hamster		
CD8a	T8 OKT8 Leu-2a	Ly-2 Lyt-2	OX8 OX9 OX10	CT6		Cytotoxic T lymphocytes (CTL)	CTL adhesion
CD8b	Leu-2b	Ly-3 Lyt-3				CTL	CTL adhesion
CD11a	LFA-1	Ly-15				Leukocytes and erythroid and myeloid stem cells	
CD11b	Mac-1 Leu-15	Ly-40	OX42			Macrophages, granulocytes, Ly-1 B cells	C3bi receptor
CD16	Leu-11a					NK cells, granulocytes	F_cR_{lo}
CD20	B1 Leu-1b					B lymphocytes, except plasma cells; follicular dendritic cells	
CD23	CD23	Ly-42				B lymphocyte subsets, follicular dendritic cells	IgE Fc receptor
CD25	TAC IL-2R	Ly-43	OX39 ART-18	5C3?		Activated T cells, B cells	IL-2 receptor
CD32	FcR_{11}	Ly-17 Lym-20				B, myeloid, and Langerhans' cells	Fc IgG2b/1 receptor
CD45	T29/35	T200	Ly-5 L-CA	OX22	201	Panleukocyte, erythroblast, and follicular dendritic cells; thymocytes; and some B cells	
CD45R	CD45R	Ly-5 B220	OX51			Pre-B and B cells, CTL, granulocytes, monocytes	
Transferrin receptor	OKT9		OX26			Activated lymphocytes, monocytes	

Hematopoietic Cell Surface Antigens Mentioned in the Text (*continued*)

Cluster of Differentiation (CD) Designation	Monoclonal Antibodies Reacting to Species-Specific Cell Surface Antigen					Distribution	Description
	Human	Mouse	Rat	Guinea Pig	Hamster		
Theta	Thy-1	Thy-1	OX7 Thy-1	154	HO-13-4	T cells, neurons, fibroblasts, and stem cells	Not present on human T cells
Ia	HLA-D (DR, DP, DQ)	H-2I (I-A, I-E)	RT1B RT1D		P51 S11 S12 S34	B cells, monocytes, activated T cells	MHC class II (Ia) antigens
Others		Lyb-3				B cells	
		Lyb-7				B and T lymphocytes, subsets of thymocytes, 15% of bone marrow cells	
		Lyb-5				B lymphocyte subset	
		ThB				B lymphocytes, thymocytes	
			OX2			Follicular dendritic cells	

SOURCES:

Murine hematopoietic cell surface antigen expression, K. L. Holmes and H. C. Morse III, eds., Immunol. Today 9:344–349, 1988.

The Lymphocyte: Structure and Function, 2d ed., J. J. Marchalonis, ed., Marcel Dekker, Inc., New York, 419 pp., 1988.

Differentiation Antigens in Lymphohemopoietic Tissues, M. Miyasaka and Z. Trnka, eds., Marcel Dekker, Inc., New York, 531 pp., 1988.

Glossary

Acute-Phase Reactants. A diverse group of proteins that progressively increase in the plasma during the first hours to days following tissue injury.

ADA (Adenosine Deaminase). An enzyme of the purine salvage pathway that catalyzes the irreversible deamination of adenosine and 2′-deoxyadenosine.

ADA Deficiency. An autosomal recessive form of severe combined immunodeficiency disease in man.

ADCC (Antibody-Dependent Cytotoxic Cells). Lymphocytes bearing Fc receptors that mediate a cellular cytotoxic reaction to target tissue with the aid of Fc-bound specific antibody.

Allogeneic. Having cells that are antigenically distinct. Refers to intraspecies genetic variations.

ALS (Anti-Lymphocyte Serum). A specially prepared serum containing

---

Sources used to compile this glossary include the following:

Blakiston's Gould Medical Dictionary, 3d ed., McGraw-Hill Book Co., New York, 1,828 pp., 1972.

Immunology, I. Roitt, J. Brostoff, and D. Male, Gower Medical Publishing, London, 316 pp., 1985.

International Dictionary of Medicine and Biology, vols. 1 and 2, John Wiley & Sons, New York, 3,200 pp., 1986.

Advanced Immunology, D. Male, B. Champion, and A. Cooke, J. B. Lippincott Co., Philadelphia, 222 pp., 1987.

antibodies against lymphocytes. Used especially to prevent rejection of transplanted organs.

Amyloidosis. Deposition of amyloid in various organs or tissues. Amyloid is a family of nonrelated beta-pleated sheet proteins that share the characteristic of extracellular fibril deposition.

ANA (Antinuclear Antibody). Any antibody capable of binding to components of the cell nucleus.

Anemic. Characterized by a reduction of erythrocytes, hemoglobin, or hematocrit to below normal levels.

Anterior Hypophysis. The anterior lobe of the pituitary gland.

Anti-DNP (Anti-Dinitrophenol). An antibody directed toward dinitrophenol. Spontaneous elevations of this antibody in the plasma have been associated with some autoimmune phenomena.

Anti-dsDNA (Anti-Double-Stranded DNA). An autoantibody frequently seen in patients with systemic lupus erythematosus and certain related connective tissue diseases.

Anti-IFN (Anti-Interferon). Antibodies directed against interferon.

Anti-SM (Anti-Smith). An autoantibody directed toward a nuclear antigen present in active chromatin. It is resistant to both deoxyribonuclease (DNase) and ribonuclease (RNase) and specific for the diagnosis of systemic lupus erythematosus.

Anti-ssDNA (Anti-Single-Stranded DNA). An autoantibody frequently seen in the plasma of people afflicted with autoimmune disease.

Ataxia Telangiectasia. An inherited disorder characterized by the onset of progressive cerebellar ataxia in infancy or childhood, oculocutaneous telangiectasia, frequent infections of the lungs and sinuses, and a propensity for the development of malignant disease. It is often associated with defects in cellular immunity and in the immunoglobulin system.

Atrophy. An acquired physiologic or pathologic local reduction in the size of a cell, tissue, organ, or region of the body.

Autoantibodies. Antibodies produced by a host to its own tissues.

Autoimmune Disease. A disease involving a humoral or cell-mediated immune attack against the tissues of one's own body.

Autoimmunity. A condition presumed to be caused by sensitization or loss of tolerance to autochthonous products of the body. A failure of the immune system to discriminate between self and nonself.

B Cells. Lymphocytes that secrete antibody and that are characterized by the presence of certain surface membrane markers, that is, immunoglobulin receptors.

Backcross. A cross between a hybrid and its parent or between a heterozygote and a homozygote.

Balanced Stock. A genetic stock that is heterozygous for closely linked genes and produces phenotypically recognized offspring types. One offspring type is used to produce the next generation and is of a predicted genotype unless crossing over has occurred between the linked loci.

Barrier-Maintained Animals. Animals, usually defined-flora or pathogen-free, placed in an environment that can serve as a barrier to the introduction of other microbes.

Basophils. A circulating polymorphonuclear leukocyte that has a small number of prominent purple or black cytoplasmic granules when stained with Romanowsky dyes. The granules contain histamine and chondroitin sulfate. In tissues it is called a mast cell.

BCGF-2 (B-Cell Growth Factor Type 2). Isolated from murine T-cell cultures this 50- to 55-kilodalton glycoprotein drives activated B cells to maturation.

BGG (Bovine Gamma-Globulin). The gamma-globulin fraction from the serum of cattle.

C3bi. Inactivated form of the split product of the third component of complement, C3b.

Calcitonin. A single-chain polypeptide hormone, apparently existing as several active fractions, secreted by the thyroid gland and the ultimobranchial bodies. A calcitoninlike substance has also been extracted from porcine adrenal glands. In hypercalcemia the hormone rapidly lowers blood calcium by inhibiting bone resorption; it also increases urinary excretion of phosphate. Called by some investigators thyrocalcitonin.

Centromere. A small body at the constriction in a chromosome where it is attached to a spindle fiber.

Chédiak-Higashi Syndrome. A human syndrome (with homologous genetic disorders in many mammalian species) with deeply staining, coarse peroxidase-positive granules in the cytoplasm of neutrophils and eosinophils. These granules are associated with albinism, hepatosplenomegaly, lymphadenopathy, and recurrent skin and pyogenic infections. When albinism is absent, the condition is called the Steinbrinck type. See the beige (*bg*) mutation of mice (p. 40).

Chemotaxis. Increased directional migration of cells, particularly in response to concentration gradients of certain chemotactic factors.

Chromosome. Intranuclear elements composed of DNA and protein that carry the hereditary factors (genes) and are present at a constant number in each species.

Class I MHC Antigens. Antigens encoded by the major histocompatibility complex (MHC) of genes that are found on all nucleated cells and are composed of two polypeptide chains of 45 and 12 kilodaltons.

Class II MHC Antigens. Also known as Ia antigens, they are found on antigen-presenting cells and are composed of two polypeptide chains of 28 and 32 kilodaltons.

cM (Centimorgans). One one-hundreth of a Morgan. A measurement of length along the chromosome having a corrected crossover frequency of 1 percent.

CMI (Cell-Mediated Immunity). A term used to refer to immune reactions that are mediated by cells rather than by antibody or other humoral factors.

Codominant. Of or pertaining to two alleles that are both expressed in a heterozygote. The phenotype is the additive function of the two alleles.

Coisogenic Strain. A congenic strain whose difference is limited to a single genetic locus.

Complement. Any one of a group of at least nine factors, designated Cl, C2, etc., that occur in the serum of normal animals and enter into various immunologic reactions. Complement is generally absorbed by combinations of antigen and antibody and, with the appropriate antibody, can lyse erythrocytes, kill or lyse bacteria, enhance phagocytosis and immune adherence, and exert other effects. Complement activity is destroyed by heating the serum at 56°C for 30 minutes.

ConA (Concanavalin A). A lectin produced by the jack bean that combines membrane glycoproteins containing alpha-glucoside or alpha-mannoside groups and serves as a mitogen for T cells.

Congenic Strain. A strain that differs from the parental inbred strain at one restricted region of the genome.

Contact Sensitivity. Also called contact hypersensitivity. An epidermal reaction characterized by eczema and caused when an antigen is applied to previously sensitized skin.

Conventional Environment. Environmental conditions where the microbial flora are unknown and uncontrolled.

Coombs' Positive Hemolytic Anemia. Anemia characterized by the attachment of anti-erythrocyte antibody and leading to sequestration of erythrocytes in the reticuloendothelial system or to complement-mediated intravascular hemolysis.

CRI (Cross-Reactive Idiotype). An idiotype that can be detected on two different antibodies is termed a cross-reactive idiotope, and the antibodies that share them are cross-reactive idiotypes.

Cyclosporine A. An immunosuppressive drug commonly used in organ transplantation. It appears to mediate its suppressive effect mainly through $CD4^+$ (helper) T cells.

Cytotoxic T Cell (Tc). A thymic-derived lymphocyte that circulates in search of a target cell displaying a determinant recognized by its receptor. It must recognize target cell antigen in association with class I major histocompatibility complex determinants.

Defined-Flora. Animals that have a microbial flora that is completely known.

Delayed-Type Hypersensitivity. A type IV immunologic reaction initiated by T lymphocytes and characterized by a delayed inflammatory response. Previous sensitization is required. Examples include response to *Mycobacterium tuberculosis* (tuberculin test) and contact dermatitis (poison ivy).

Dendritic Cells. Antigen-presenting cells present in lymph nodes, spleen, and at low levels in blood that are particularly active in stimulating T cells.

DES (Diethylstilbestrol). A nonsteroid estrogen used as a substitute for the natural estrogenic hormones, it is more readily absorbed from the alimentary canal than most of the natural hormones and hence is suitable for oral use.

Diabetes Mellitus. An inherited chronic disorder of carbohydrate metabolism caused by a disturbance of the normal insulin mechanism and characterized by hyperglycemia, glycosuria, and alterations of protein and fat metabolism. It produces polyuria, polydipsia, weight loss, ketosis, acidosis, and coma.

DNFB (2,4-Dinitro-l-Fluorobenzene). A chemical that, when applied topically to skin, sensitizes the host so that a second application causes contact sensitivity.

Dysgenesis. Abnormal development of anything, usually of an organ or individual. Impairment or loss of the ability to procreate.

EAE (Experimental Allergic Encephalomyelitis). A disease state induced by the inoculation of allogeneic spinal cord that is characterized by inflammation of the brain and spinal cord.

Erythropoiesis. The differentiation and proliferation of erythrocytes.

FACS (Fluorescence-Activated Cell Sorting). A procedure for mechanically separating cells based on the presence of membrane-bound fluorescent dyes. Commonly, the dye is attached to a specific antibody, which directs it to a particular plasma membrane determinant.

Fc Receptors. Membrane receptors in cells capable of binding the Fc portion of immunoglobulin molecules.

FTS (Facteur Thymique Serique). A nine-amino-acid residue thymic hormone; also called thymulin.

F Number. Indicates the number of filial or brother × sister generations.

GA (Gallic Acid). 3,4,5-Trihydroxybenzoic acid, $C_7H_6O_5$, formerly used internally as an astringent; esters of the acid are used as antioxidants.

Gene. The biologic unit of heredity. It is self-reproducing and located at a specific location (locus) on a specific chromosome.

Genotype. The genetic material inherited from parents; not all of it is necessarily expressed in the individual (cf. phenotype).

Germfree Environment. An environment free of all known microbes.

Germfree Isolator. An apparatus used to maintain animals in a germfree environment.

GIF (Glucocorticoid Increasing Factor). A substance made by certain lymphocytes that exerts its stimulatory effect through the pituitary gland.

Glomerulonephritis. An acute or chronic, usually bilateral, diffuse nonsuppurative inflammatory kidney disease primarily affecting the glomeruli. It is characterized by proteinuria; cylindruria; hematuria; and often edema, hypertension, and nitrogen retention.

Glucocorticoid. An adrenal cortex hormone that affects the metabolism of glucose. Any related natural or synthetic substance that functions similarly. In pharmacologic doses, these substances have immunosuppressive effects.

Gnotobiotic. Germfree animals or germfree animals into which a defined microflora is introduced. The defined flora should be few in number and nonpathogenic.

gp70 (Glycoprotein 70). A 70,000-dalton glycoprotein found on the envelope of murine leukemia viruses.

GM-CSF (Granulocyte-Macrophage Colony Stimulating Factor). A glycoprotein hormone that stimulates myeloid progenitor cells to produce neutrophils, macrophages, and eosinophils.

GVH (Graft-Versus-Host Reactions). An immunologic reaction characterized by donor (grafted) lymphocytes attacking host cells based on the recognition of histocompatibility antigens on host cells not present on the donor lymphocytes and the inability of host lymphocytes to reject the donor cells.

Haplotype. A set of genetic determinants located on a single chromosome.

Hashimoto's Thyroiditis. A chronic autoimmune disease of the thyroid, more common in women, that presents as thyroid gland enlargement and hypothyroidism. The gland is densely infiltrated with lymphocytes and is ultimately destroyed.

Hematocrit. The percentage of the whole-blood volume occupied by erythrocytes after centrifugation.

Hemizygous. The presence of only one of a pair of genes that influence the determination of a specific trait. For example, a male has few alleles on his Y chromosome; therefore, most of the alleles on his X chromosome are expressed whether they are dominant or recessive.

Henoch-Schonlein Purpura. A disease, primarily of young children, consisting of small-vessel vasculitis seen as raised red spots on the skin, glomerulonephritis, abdominal pain, fever, and arthritis. It is thought to be a form of allergic vasculitis.

HEPA (High-Efficiency Particulate Air Filter). A filter capable of retaining 99.97 percent of a 0.3-μm diameter monodispersed aerosol.

Heterozygosity. Having dissimilar alleles in one or more pairs of gene loci.

Heterozygote. An individual that has different alleles in the corresponding loci of homologous chromosomes and does not breed true.

HGG (Human Gamma-Globulin). The gamma-globulin fraction of human serum. An extract of serum from hyperimmune humans that is used as passive immunotherapy in patients exposed to an infectious disease, for example, rabies immune globulin.

Histamine. 4-(2-Aminoethyl)-imidazole, $C_5H_9N_3$, an amine occurring as a decomposition product of histidine that stimulates visceral muscles; dilates capillaries; and stimulates salivary, pancreatic, and gastric secretions. It is found in the granules of the basophils and mast cells responsible for anaphylactic reactions.

Holandric Inheritance. Transmission of a trait via the male lineage; transmitted through genes on the Y chromosome.

Homozygote. An individual that has identical alleles in the corresponding loci of homologous chromosomes and breeds true.

Hybrid. The offspring of a cross between two inbred strains.

Hypergammaglobulinemia. The increased concentration of immunoglobulins in the blood seen in a wide variety of clinical disorders.

Hyperplasia. Excessive formation of tissue; an increase in the size of a tissue or organ caused by an increase in the number of cells.

Ia-Associated Antigen. A surface antigen of mouse cells, such as B lymphocytes and macrophages, that is determined by the *I-A* region of the major histocompatibility complex.

Idiotope. A single antigenic determinant on an antibody V region.

Idiotype. The antigenic characteristic of the V region of an antibody.

IL-1 (Interleukin-1). An acute-phase reactant synthesized by many cell types, including monocytes and lymphocytes. This hormone has many effects, including the activation of resting T cells, the promotion of synthesis of other lymphokines, and the activation of macrophages and endothelial cells.

IL-2 (Interleukin-2). A growth factor for activated T cells, this hormone is derived from lymphocytes and promotes the synthesis of other lymphokines.

IL-3 (Interleukin-3). A multilineage colony stimulating factor released by T cells that appears to act synergistically with granulocyte-macrophage colony stimulating factor to stimulate hematopoiesis. It is also a growth factor for mast cells.

IL-4 (Interleukin-4). Also known as B-cell stimulating factor type 1, this is a growth factor for B cells and induces class II major histocompatibility complex antigen expression on their surfaces. This hormone also enhances the cytolytic activity of cytotoxic T cells and is a mast cell growth factor.

IL-6 (Interleukin-6). Also known as B-cell stimulating factor type 2 or B-cell differentiation factor, it induces the differentiation of activated B cells into immunoglobulin-secreting plasma cells.

Immune Complex. The product of an antigen–antibody reaction that can also contain components of the complement system.

Immunodeficiency. The inability of the host to eliminate or neutralize foreign substances.

Immunologic Memory. A phenomenon characterized by the presence in the body of an expanded set of clonally derived antigen-specific lymphocytes that can be rapidly recruited to produce an augmented immune response on subsequent exposure to the specific antigen.

Immunoglobulins. Proteins of animal origin with known antibody activity or a protein related by chemical structure and hence antigenic specificity. They can be found in plasma, urine, spinal fluid, and other body tissues and fluids and include such proteins as myeloma and Bence Jones protein.

Inbreed. To mate brother × sister each generation.

Inbred Strain. A strain produced by at least 20 generations of brother × sister inbreeding.

Incisor Teeth. The two cutting teeth nearest the midline in each quadrant of the dentition. Also called dentes incisive.

Intercrossing. The mating of diploid individuals who are both heterozygous for one or more of the same alleles at given loci.

Interferons. Proteins that are formed by animal cells in the presence of a virus, that prevent viral reproduction, and that are capable of inducing resistance to a variety of viruses in fresh cells of the same animal species. There are three classes: alpha, beta, and gamma. Alpha interferon (IFNα) is made by lymphocytes and macrophages and is 18–20 kilodaltons in size. Beta interferon (IFNβ) is synthesized by fibroblasts and epithelial cells. Alpha and beta interferons were once called type 1 interferon. Gamma interferon (IFNγ), also called type 2 interferon, is synthesized by lymphocytes. All three interferons can be induced during viral infection. They have antiviral and antiproliferative effects, and all induce class I major histocompatibility complex antigens.

Islets of Langerhans. Islets of the pancreas containing hormone-producing cells: beta cells produce insulin; alpha cells produce glucagon.

Isotypes. Genetic variation within a family of proteins or peptides such that every member of a species has each isotype of the family represented in its genome (e.g., immunoglobulin classes).

Kappa (κ) Chain. One of the immunoglobulin light-chain isotypes.
Ketosis. A condition in which ketones are present in the body in excessive amounts. The acidosis of diabetes mellitus.

Lambda (λ) Chain. One of the immunoglobulin light-chain isotypes.
Laminar Flow. Air flow characterized by the absence of turbulence.
Langerhans' Cells. Stellate cells of the mammalian epidermis and dermis presumed to belong to the melanocyte series, revealed by gold impregnation, and containing nonmelanized disklike organelles. They are antigen-presenting cells that emigrate to the local lymph nodes to become dendritic cells. They are particularly active at presenting antigen to T cells.
Lentivirus. One member of the virus family Retroviridae. Many species have the ability to infect animals, including man, and have long incubation periods. These viruses usually are not oncogenic.
Linked Genes. In eukaryotic cells, genes are said to be linked if they are carried on the same chromosome, that is, do not segregate randomly.
Lipid A. The endotoxic component of lipopolysaccharide.
LPS (Lipopolysaccharide). A compound of a lipid with a polysaccharide. A component of gram-negative bacteria that is liberated on bacterial death and might be responsible for shock; also called endotoxin.
Lymphadenopathy. Lymph node enlargement.
Lymphoblasts. A blast cell, considered a precursor or early form of a lymphocyte.
Lymphocyte Activating Factor. Also called interleukin-1.
Lymphokines. A generic term for molecules other than antibodies that are involved in signaling between cells of the immune system and are produced by lymphocytes (cf. interleukins).
Lymphomas. Any neoplasm, usually malignant, of the lymphatic tissues.
Lymphopenia. Lymphocytopenia. A reduction in the number of circulating lymphocytes.
Lymphopoietic. Lymphocytopoiesis. Production of lymphocytes.

Macrophages. A phagocytic cell that is found in many tissues, that is derived from a blood monocyte, and that has an important role in host defense mechanisms.
MALT (Mucosal-Associated Lymphoid Tissue). Generic term for lymphoid tissue associated with the mucosa of the gastrointestinal tract, bronchial tree, urinary tract, and other systems.
Mast Cell. A small cell similar in appearance to a basophil and found associated with mucosal epithelial cells. These cells are dependent on T cells for proliferation, and they contain cytoplasmic granules laiden with heparin, slow reactive substance of anaphylaxis, and eosinophil chemo-

tactic factor of anaphylaxis, which are released when antigen binds to membrane-bound IgE.

Melanocyte. The epidermal cell that synthesizes melanin.

Membrane Attack Complex. A multimolecular complex formed by the activation of the terminal components of the complement pathway and responsible for cytolysis.

MHC (Major Histocompatibility Complex). A genetic region found in all mammals whose products are primarily responsible for the rapid rejection of grafts between individuals and that function in signaling between lymphocytes and cells expressing antigen.

MHV (Mouse Hepatitis Virus). A single-stranded RNA virus of the family Coronaviridae, genus *Coronavirus*. MHV infections are usually subclinical in immunocompetent mice but can cause severe illness and death in immunodeficient mice.

Mitogens. Substances that cause cells, particularly lymphocytes, to undergo cell division.

MLC (Mixed Lymphocyte Culture). The proliferative assay system for T-cell recognition of foreign antigen on allogeneic cells, in which response is measured by DNA synthesis (intercollation of radiolabeled nucleic acid).

MLR (Mixed Lymphocyte Reaction). See MLC.

Monoclonal Antibodies. Antibodies arising from a single clone of B lymphocytes.

MuLV (Murine Leukemia Virus). An RNA virus of the family Retroviridae. An endogenous virus of mice that causes leukemias and related malignancies in some mouse strains. Strain susceptibility is influenced by a number of genes, including *Fv-1, Fv-2, In, nu, hr,* and *Ir*.

Mutations. A change in the characteristics of an organism produced by an alteration in the hereditary material. The alteration in the germplasm might involve an addition of one or more complete sets of chromosomes; the addition or loss of a whole chromosome; or some change within a chromosome, ranging from a gross rearrangement, loss, or addition of a larger or smaller section to a minute rearrangement or change at a single locus. A small change at a single locus is called a gene mutation or, more simply, mutation in the restricted use of the term.

Myelopoiesis. The process of formation and development of blood cells in the bone marrow.

N. Indicates the number of backcrosses made to an inbred strain.

Necrosis. The pathologic death of a cell or group of cells.

NK (Natural Killer) Cells. A group of lymphocytes with the intrinsic ability to recognize and destroy some virally infected cells and some tumor cells.

Nephritogenic. Giving rise to inflammation of the kidney.

NTAs (Natural Thymocytotoxic Antibodies). Autoantibodies directed against determinants present on thymic-derived lymphocytes, as defined by complement-dependent lysis of target thymocytes.

Oncornavirus. Any of the various tumor-producing viruses of the family Retroviridae. Also called oncovirus.

Osteoclast. A multinuclear giant cell responsible for bone absorption and destruction.

Ovarian Transplantation. A technique of placing a compatible ovary in the ovarian capsule (from which the original ovary has been removed) of a selected host.

Osteopetrosis. A heterogeneous group of hereditary disorders that share generalized sclerosis and fragility of the skeleton and elevated serum acid phosphatase.

Osteosclerosis. An abnormal thickening of bone.

P-15E. A transmembrane portion of the envelope of certain oncoviruses thought to be responsible for viral-associated immunosuppression.

Parabiosis. The experimental surgical fusing together of two individuals or embryos so that the effects of one partner on the other can be studied.

Parathyroid Hormone. A polypeptide hormone in parathyroid glands that regulates blood calcium levels.

Pathogen-Free Environment. An environment free of all known pathogenic microorganisms, tested pathogens, or both.

PBB (Polybrominated Biphenyl). Any of a series of stable compounds in which the hydrogen atoms in biphenyls are replaced by bromine. These, along with polychlorinated biphenyls, are serious toxic pollutants in the environment.

PBLs (Peripheral Blood Lymphocytes). Circulating lymphocytes found in blood.

Penetrance. The percentage of organisms with a given genetic constitution that show the corresponding hereditary characteristic.

Peyer's Patches. Specialized lymphoid tissue found in the intestinal tract.

PFC (Plaque-Forming Cell). An antibody-secreting B cell that can be recognized by the production of a hemolytic plaque.

PG (Propylgallate). 3,4,5-Trihydroxybenzoic acid propyl ester, an antioxidant.

PHA (Phytohemagglutinin). A lectin isolated from red kidney beans that is mitogenic for T cells.

Phagocytosis. Ingestion of foreign or other particles by certain cells.

Phenotype. The expressed characteristics of an individual (cf. genotype).

Pleiotropic. The occurrence of multiple effects produced by a given gene.

PMN (Polymorphonuclear Leukocyte). The mature neutrophil leukocyte, so-called because of its segmented and irregularly shaped nucleus.

Pneumonitis. Inflammation of the lungs in which the exudate and inflammatory cell incursion is primarily interstitial.

Poly(A-U) (Polyadenylic-Polyuridylic Acid). A synthetic RNA polymer.

Polyarteritis Nodosa. A systemic disease characterized by widespread inflammation of small- and medium-sized arteries in which some of the foci are nodular. Complications of the process, such as thrombosis, lead to retrogressive changes in the tissues and organs supplied by the affected vessels with a correspondingly diverse array of symptoms and signs. It is also called periarteritis nodosa or disseminated necrotizing periarteritis.

Poly (I-C). One of the synthetic RNA polymers that induce the local production of interferon. It has been used experimentally in the topical treatment of acute herpesvirus infection of the cornea.

Prokaryotic. Pertaining to a unicellular organism that has a single chromosome, lacks a nuclear membrane, and usually has a rigid peptidoglycan wall.

PGE2 (Prostaglandin E2). An unsaturated fatty acid 20 carbons in length with an internal cyclopentane ring. It causes vasodilatation, inhibits gastric secretion, induces labor and abortion, and is immunosuppressive. A derivative of arachidonic acid.

Protein A. A cell wall component of certain strains of staphylococci that binds to a site in the Fc region of most IgG isotypes.

PWM (Pokeweed Mitogen). A mitogen for B cells.

RCS (Reticulum-Cell Sarcomas). A malignant tumor in which the predominant cell type is an anaplastic reticulum cell. Multinucleated cells also occur. Also called histiocytic sarcoma.

Recombinant Inbred. A strain formed by crossing two inbred strains, followed by 20 or more generations of brother × sister mating.

Reed-Sternberg Cell. A large binucleated or multinucleated cell, 15–45 μm in diameter, that is derived from the reticuloendothelial system. It is a distinctive giant cell found in all Hodgkins' lymphomas.

Repulsion. The occurrence of two linked loci, each in the heterozygous state, with the mutant alleles on different homologous chromosomes.

RES (Reticuloendothelial System). A diffuse system of phagocytic cells, many of which are derived from bone marrow stem cells. It is associated with the connective tissue framework of the liver, spleen, lymph nodes, and other serous cavities. The site of renewal for endothelial cells is thought to be existing endothelial cells rather than bone marrow.

Retroviral. Relating to retroviruses, RNA viruses belonging to the family Retroviridae that are characterized by the presence of a reverse transcriptase (RNA-dependent DNA polymerase) enzyme.

RFs (Rheumatoid Factors). A group of autoantibodies that are directed against the Fc fragment of the heavy chain of IgG.

RFLP (Restriction Fragment Length Polymorphism). DNA fragments of various sizes resulting from the action of endonucleases cleaving DNA at specific sites (restriction enzymes).

Semidominant. Incomplete dominance.

Serotonin. 5-Hydroxytryptamine, $C_{10}H_{12}N_{20}$, which is present in many tissues, especially blood and nervous tissue. It stimulates a variety of smooth muscles and nerves and is postulated to function as a neurotransmitter.

Serum Alkaline Phosphatase. Serum levels of a specific phosphatase enzyme found in various tissues. Elevation of this enzyme in the serum usually indicates hepatobiliary disease in mammals.

SAA (Serum Precursor Amyloid). An alpha-globulin that displays acute-phase increases and is believed to be the precursor of the amyloid A fibrils present in secondary amyloidosis.

SDS Gel Electrophoresis. A form of polyacrylamide gel electrophoresis conducted with sodium dodecyl sulfate (SDS) in the buffer.

SPF (Specific-Pathogen-Free). Free of specifically defined pathogenic microorganisms.

Splenomegaly. Increased spleen size; it is seen in a number of parasitic infections, hemolytic anemias, and lymphomas and is also measured in the Simonson assay for graft-versus-host reactions.

SRBC (Sheep Red Blood Cell). A T-cell-dependent target antigen often used in hemolytic plaque assays of immune responsiveness.

Stem Cell. Pluripotent cells that can serve as progenitor cells for the lymphoid lineage or the myeloid lineage or both (hemopoietic stem cell).

SV40 (Simian Virus 40). A papovavirus commonly seen in tissues of Old World monkeys without being associated with disease.

Synergistic. An agent that increases the action or effectiveness of another agent when combined with it.

Syngeneic. Individuals or tissues that have identical genotypes, for example, animals of the same inbred strain.

SLE (Systemic Lupus Erythematosus). An autoimmune disease of humans usually involving antinuclear antibodies and characterized by skin rash, hematologic alterations, and glomerulonephritis.

TCDD (Tetrachlorodibenzo-*p*-Dioxin). One member of a group commonly referred to as dioxins that is widespread in the environment and is immunosuppressive.

T-Cell- or Thymic-Derived Lymphocytes. One of the two major classes of lymphocytes with important immune regulatory and effector functions.

T cells must pass through the thymus during development, and they carry certain characteristic surface markers such as theta and the CD3 antigens.

Tdt (Terminal Deoxynucleotidyl Transferase). An enzyme present in pre-T or thymic stem cells. It is present in cortical but lost in medullary and peripheral cells.

Th Cell or Helper T Lymphocyte. A functional subclass of T cells that can help generate cytotoxic T cells and cooperate with B cells in the production of an antibody response. Helper T cells usually recognize antigen in association with class II major histocompatibility complex molecules. Currently defined using CD4 markers.

Thymulin. An 867-dalton peptide isolated from thymus and serum that appears to be responsible for T-cell differentiation. Also known as facteur thymique serique (FTS).

Thymus. An organ found in the thoracic or cervical regions of mammals that is composed of lymphatic tissue in which minute concentric bodies, the remnants of epithelial structures, or thymic corpuscles are found. This organ is necessary for the development of thymic-derived lymphocytes and is the source of several hormones involved in T-cell maturation, for example, thymosin, thymopoietin, thymulin, and thymocyte humoral factor.

Thymus-Dependent Antigen. An antigen that requires an immune response from thymic-derived lymphocytes in order to elicit an immune response from B lymphocytes.

Thymus-Independent Antigen. An antigen that does not require the participation of T lymphocytes to elicit an immune response in B cells.

Ti or TCR (T-Cell Antigen Receptor). The antigen receptor of T cells composed of two polypeptide chains and closely associated with the T3 surface membrane molecules.

TLI (Total Lymphoid Irradiation). X-irradiation directed toward lymphoid organs throughout the body.

Tolerance. A state of specific immunologic unresponsiveness.

Ts Cell or Suppressor T Lymphocyte. A subpopulation(s) of T cells that acts to reduce the immune responses of other T or B cells. Suppression can be antigen specific, idiotype specific, or nonspecific under different circumstances. At present, cells with this function cannot be identified using one marker, although many appear to carry the CD8 molecule.

Type 1 Antigens. T-cell-independent (TI) antigens that can stimulate both $Lyb\text{-}5^+$ and $Lyb\text{-}5^-$ B cells in the mouse. TI type 2 antigens only stimulate $Lyb\text{-}5^+$ B cells.

UVR (Ultraviolet Radiation). Radiation with a wavelength shorter than that of the violet end of the visible spectrum and longer than that of x rays (from 180 to 390 nm). Exposure to UVR can be associated with tissue damage, malignant transformation, or immunosuppression.

Vibrissae. One of the hairs in the vestibule of the nose. One of the long, coarse hairs on the face of certain animals; "whiskers."

Viral-Antibody-Free. Free of circulating anti-viral antibodies to specific viruses, usually pathogenic viruses.

Vitiligo. A condition of the skin characterized by a failure to form melanin, with patches of depigmentation that often have a hyperpigmented border and enlarge slowly.

Wild Type. Of or relating to a genetic locus or an allele that specifies a phenotype that predominates in natural populations or that is designated as normal.

Xenografts. A transplant from one species to another; sometimes used to indicate a wider genetic or species disparity than in a heterograft.

Index

A

Adaptive immune system/mechanisms, 2–10, 224
Adenosine deaminase, 85, 217; *see also* Severe combined immunodeficiency
β-Adrenergic substances, 10
Allergic contact dermatitis, 81, 220
Alopecia, 57, 116; *see also* Mouse mutants, *nu*; Rat mutants, *rnu*
Amyloidosis, 116, 119, 122, 218
Anemia
  autoimmune hemolytic, 88, 97, 105–106, 112, 220
  defined, 218
  macrocytic, 82–83
  in osteopetrotic mutations, 51, 68
Antibiotics, and immunosuppression, 12
Antibodies
  anti-class II MHC, 132
  anti-DNP, 135, 218
  antihistone, 115
  antinuclear, 50, 54, 61, 70, 88, 94, 99, 109, 114, 115, 122, 218
  B cell production of, 3
  deficiency in LPS-induced adjuvant response to, 63
  enhanced production of, 142
  function of, 3–4, 10
  genetic mechanisms in diversification of, 6–7
  T-cell-dependent responses, 140–141, 142
  T-cell independent responses, 140–141
  thymocyte-binding, 88
  viral inhibition of, 13
  see also Autoantibodies; Immunoglobulins
Antibody-dependent cytotoxic cells, 10, 217
Antigens
  Brucella, 50
  defective responses to, 115, 116, 135, 136
  designations, 2
  IaW39, 86
  I-E surface, 101
  monoclonal antibodies reacting to, 213–215
  resistance to tolerance induction by, 119
  retroviral gp70, 108, 112
  T-cell dependent, 116, 135–137
  T-cell independent, 116, 230
  thymus-dependent, 85, 115, 230
  thymus-independent, 85, 230
  thyroid-stimulating hormone receptor (TSH), 14
  see also Major histocompatibility complex antigens

Anti-lymphocyte serum, 145, 217–218
Anti-theta serum, 145
Arteritis, 60, 99, 114, 121, 228
Arthritis, 14, 60
Ataxia telangiectasia, 84–85, 218
Autoantibodies
　against islet antigens, 101
　anti-DNA, 50, 88, 91, 99, 105, 106, 108, 112–114
　anti-double-stranded DNA, 54, 61, 66, 90, 91, 105, 108–109, 218
　anti-erythrocyte, 105–106, 108
　anti-gp70, 105, 107, 108, 110, 121
　anti-insulin, 45, 104
　anti-islet cell cytoplasmic antigens, 45
　anti-pancreatic beta cells, 44–45
　anti-p73, 104
　anti-ribonucleic protein (RNP), 99
　anti-RNA, 110
　anti-single-stranded DNA, 61, 90, 91, 105, 106, 108, 121, 218
　anti-Smith, 61, 99, 218
　anti-thymocyte, 66, 105
　characteristic of autoimmune disease, 14
　cross-reactive idiotype families, 113
　defined, 218
　genes controlling, 108
　natural cytotoxic antibodies (NCA), 106
　reduced levels of, 91
　thymus-binding, 54, 227
Autoimmunity/autoimmune disease
　acceleration of, 87–89, 91, 93–94, 97, 112
　autoantibodies characteristic of, 14
　autoimmune polyendocrine disease, 14
　B cell depletion and, 94
　correction of, 107–108
　defined, 2, 218
　early-onset lymphoid hyperplasia with, 52
　environmental factors in, 13–14
　estrogen therapy and, 99
　experimental autoimmune encephalomyelitis (EAE), 95, 117–118, 120, 221
　genes controlling susceptibility to, 105, 161
　hemolytic anemia, 105
　induced by MHV, 162
　Ly-1$^+$ B cell population and, 66
　modification of natural history, 86–87
　orchiectomy and, 88
　progressive autoimmune lymphoproliferative disease, 94
　spontaneous, 13–14, 70, 101, 105
　systemic lupus erythematosus, 54, 99, 105, 108–111, 114–115, 121, 122, 135
　against thymic hormones, 45
　without lymphoid hyperplasia, 99
　Y-linked, 38, 87–89, 91, 93–94, 97

B

B cell defects
　absence of mature cells, 92–94
　atrophy, 101
　and autoimmunity, 111, 113
　blocked complement receptors on, 141
　decreased numbers of cells, 47
　diminished response to type I antigens, 86
　functional lesions, 85–86
　hyperactivity, 107, 111, 118–119
　in Ia expression, 63
　in immunoglobulin expression, 63, 86, 107
　inability to proliferate in presence of LPS, 63
　increased function in mutants, 45, 126
　ionizing radiation and, 145
　Ly-1$^+$ population, 66, 86, 107
　Lyb-3 expression, 86
　in Lyb-5 subpopulation, 86, 92, 107
　maturation factors, 98
　polyclonal activation, 66
　production of pathogenic IgG anti-DNA, 109
　proliferative response impairment, 63, 66, 78
　resistance to induction of immunologic tolerance, 107
　in response to T-cell replacing factors, 98
　sIg$^+$ population, 86
　thymic proliferation, 119
B cells
　activation of, 9, 163
　differentiation of, 3, 4, 6–7, 9, 116
　function of, 3, 4, 8, 9, 218
　immunoglobin secretion, 3, 5, 7
　interaction with T cells, 3, 7, 9, 92, 98, 109, 112, 116, 119
　maturation of, 91
　proliferation of, 3, 4, 6–7, 9
　see also Antibodies
Babesia rodhaini, 143

Bacille Calmette-Guérin (BCG), 127
Bacillus enteritidis, resistance and susceptibility to, 94–96
Bacillus piliformis, 143
β cells
  destruction of, 102, 133
  necrosis of, 44
  unrestricted hyperplasia of, 44, 72
Blastomyces dermatitidis, 163
Bone marrow, model of functional relationships with spleen and thymus, 93
Bone marrow defects
  decreased cells, 76
  depression of granulocyte macrophage precursors, 141
  hyperplasia, 105
  increased myelopoiesis, 66
  lack of marrow cavities, 129
  in long bones, 52, 74
  lymphocyte deficiencies, 77
  plasma cell deficiencies, 77
  preleukemic cells, 71
  reduced erythropoiesis, 66
Bone marrow transplantation, correction of defects with, 42, 52, 69, 78, 83, 85, 107–108, 111, 124, 125
Bordetella pertussis, induction of histamine sensitivity by, 118, 120
Bronchopneumonia, 60
Butylatedhydroxyanisole, 141–142

C

Cadmium, immunosuppressive effects of, 11, 141
Calcitonin, 10, 51–52, 68, 74, 76, 219
Candida albicans, 137
Cell-mediated immunity, depressed, 45, 47, 49, 58, 66, 72, 110, 111, 141, 144
Cell surface antigens, monoclonal antibodies reacting to, 2, 213–215
Chédiak-Higashi syndrome, 15, 41, 218
Chemotaxis
  decreased, in mutants, 4, 41–42
  defined, 219
  drug depression of, 12
  in peritoneal macrophages, 68
Chromosome maps/gene markers
  humans, 6
  information sources on linkages, 16
  mouse, 6, 16–31

rat, 32–33
restriction fragment length polymorphism map, 86
Coccidioides immitis, 163
Complement system
  activation of, 3, 4, 6, 9–10, 55, 90
  C1, 90
  C2, 90, 134–135, 136
  C3, 50, 115, 135–136, 137, 219
  C4, 89–90, 129–130, 135, 136–138
  C5, 54, 137
  C6, 139
  characteristics and functions of, 4, 9, 220
  deficiencies, 129–130, 134–138, 139, 141
  pathways (cascades), 4
  regulator of complement activation (RCA), 90
  role of classical pathway during infection, 137–138
  S-region-linked genes controlling murine C4, 89–90
Complementarity-determining regions (CDRs), 109–110
ConA, impaired response to, 47, 84, 91, 127, 141, 145, 220
Congenic mouse strains
  A.SW, 117
  BXSB-*xid*, 94
  B10.D2/oSnJ, 55
  C.B-*Igh-1*^{b} (CB17), 77–80
  C57BL/KsJ.B6-*H-2*^{b}, 73
  NOD.NON-*H-2*, 100, 101
  NZB.CB/N-*xid*, 106–107
  SJL/J-*lpr*, 117
Congenic strains, defined, 16, 220
Conjunctivitis, 128
Corynebacterium parvum, 119, 127–128
Coxsackievirus B4, 14
Cryptococcus neoformans, 137
Cytolysis, 4
Cytomegalovirus, 143

D

Demyelinating disease, susceptibility to, 117, 120, 162, 163
Dendritic cells
  follicular, antigen retention by, 126
  function of, 3, 8, 221
  Thy-1$^+$ epidermal cells, lack of, 77
  see also Langerhans' cells

Dextran, failure to respond to, 85
Diabetes mellitus
  characteristics of, 221
  and congenital malformations, 134
  environmental modifiers of, 101–102
  insulin-dependent, 100–101, 130–134
  modifiers of severity, 44, 102
  mouse models, 17, 37, 43–45, 100–103
  non-insulin-dependent, 44, 45
  non-obese, 100–104, 132
  orchiectomy and, 101
  ovariectomy and, 101
  prevention of, 102, 103, 132
  rat models, 130–134
  resistance to, 44, 100, 131
  spontaneous autoimmune, 100–104
  transfer of, 102
  viruses associated with, 13–14
Diabetogenic encephalomyocarditis virus, 73
Diet
  autoclaved, 128, 144
  contamination through, 11, 103, 144, 152
  for diabetic mice, 103
  and immunosuppression, 11, 103, 144
  magnesium-deficient, 144
  restriction, 45, 73
  soft, 69, 75, 76, 124, 125, 129
  water, 152–153
Diethylstilbestrol, 140–141, 221
1,25-Dihydroxyvitamin D, 10, 68, 74, 124, 125
2,4-Dinitro-1-fluorobenzene (DNFB), contact sensitivity to, 81, 221
Dinitrophenol, increased PFC response to, 115

E

Ectromelia virus, 143, 147, 161, 162
Encephalomyocarditis virus, 14, 163
Endogenous ecotropic virus, 121
Endoparasites, 10
β-Endorphin, 10
Endotoxin, see Lipopolysaccharide
Eperythrozoon coccocides, 12
Escherichia coli, susceptibility to, 63
Estrogen, 10, 99
Experimental allergic encephalomyelitis (EAE), 95, 117–118, 120, 162, 221

F

Facteur thymique serique, 50, 122, 221
Factor H expression, 90
Feline leukemia virus (FeLV), 13
Flaviviruses, 161–162, 163

G

Gallic acid, 141–142, 221
Gamma-globulin, tolerance to, 80
Generalized arterial disease, 60
Generalized lymphoproliferative disease, 37, 52–54, 90
Giardia muris, 164
Glomerulonephritis, 60, 66, 88, 99, 107, 109, 112–114, 121, 122, 222
Glucagon, 14
Glucan, 146
Glucocorticoid increasing factor, 142–143, 222
Glucocorticoids, 10, 142–143, 222
Glucose intolerance, 101, 104
Glycosuria, 101
Glycosylation pathway, deficiency in, 86
Graft-versus-host reaction
  defined, 222
  impaired, 46, 49, 58, 110, 119, 127, 142, 143
Gram-negative bacteria, 62
Granulocyte-macrophage colony-stimulating factor, 64, 222
Gross virus, 13
Growth hormone, 10, 14, 46
Growth retardation, 46, 49
Guinea pig mutants, see Table 2-1, p. 39
Guinea pigs, monoclonal antibodies reacting to cell surface antigens in, 213–215

H

Halothane, 12
Hamster mutants, 1, 14; see also Table 2-1, p. 39
Hamsters
  immunoglobulin subclasses, 5
  monoclonal antibodies reacting to cell surface antigens in, 213–215
Haptenated-Ficoll, failure to respond to, 85
Hemobartonella muris, 12
Henoch-Schönlein purpura, 135, 222
Heparin, 10

Hereford cows, partially albino, 41
Herpes simplex virus (HSV), 63, 143, 147, 161, 163
Histamine, 10, 118, 120, 223
Hodgkin's disease, 117
Hormones
  and host defense function, 10
  and murine lupus erythromatosus, 111
Human chromosomes, 6
Human severe combined immune deficiency, model for, 77–79
Human T cell lymphotrophic virus type I (HTLV-I), 13
Humoral immunity, deficiencies in, 66, 144
Husbandry practices
  bedding, 153
  caging or housing system, 112, 120, 148, 149, 151
  cesarean derivation, 128
  conventional environment, 14
  defined-flora, 14, 71, 219, 221
  disease surveillance, 154
  environmental conditions, 14, 59, 151–153
  filter hoods, 128, 149
  general considerations, 14–15, 69, 148–149, 151
  germ-free isolation, 14, 67, 71, 75, 102, 127, 128, 133, 150–151, 222
  HEPA-filtered laminar-air-flow systems, 149–150, 151, 223
  and immunosuppression, 12, 50, 55
  infection control in commercial stock, 153–154
  insulin treatment for diabetic rodents, 203, 131
  laminar flow, 133
  pathogen-free environment, 1, 14, 43, 51, 71, 76, 79, 96, 108, 112, 121, 125, 127–129, 133, 139, 149, 219, 227, 229
  prevention of fighting among rodents, 120, 154
  retirement and replacement of breeders, 62
  single-user, SPF, barrier-protected room, 79, 128
  strict isolation, 14
  temperature conditions, 59, 151–152
  transfer of animals between facilities, 153–154
  see also Diet

Hybrid mouse strains (see also Table 2-1, p. 38)
  CBA/Ca × BXSB/Mp F_1, 94
  CBA/N × BXSB/Mp F_1, 94
  C3HeB/FeJ × C57BL/6J F_1, 159
  C3HeB/FeLe × C57BL/6J F_1, 84
  C57BL/6J × C3H/H F_1, 34
  C57BL/6J × C3H/HeJ F_1, 34
  C57BL/6J × C3H/HeSnJ F_1, 59, 160
  C57BL/6J × C3HeB/FeJLe-a/a F_1 (B6C3Fe-a/a F_1), 47, 48, 74, 75, 159, 160
  C57BL/6J × SB/Le, 93
  nomenclature, 34
  NZB × NZW F_1 (BWF$_1$), 86–87, 99, 105, 121, 122
  PN/Sw × NZB/Sw F_1 or NZB/Sw × PN/Sw F_1, 114
  SJL/J × BALB/c F_1, 120
  WB/ReJ × C57BL/6J (WBB6F), 82
Hydrocephalus, 76
Hypercholesterolemia, 101
Hypergammaglobulinemia, 135, 223
Hyperglycemia, 43, 72, 101, 131–132, 134
Hyperimmunoglobulinemia, 54, 88, 94
Hyperinsulinemia, 43, 44, 72
Hyperphagia, 43, 72
Hypersensitivity reactions
  contact, 81
  delayed-type, 12, 84, 119, 141, 142, 161, 221
  enhanced responsiveness, 81
  impaired, 119
  vasoactivity amine, 118
Hypocalcemia, 68, 129
Hypophosphatemia, 68, 74, 129
Hypophysis, 49, 218
Hypothermia, 59, 63

I

Immune complex
  circulating, increases in, 121
  defined, 224
  deposition in kidneys, 44, 66, 73
  glomerulonephritis, 60, 88, 97, 109
  reduced levels of, 91
  retroviral gp70, 61, 99
  serum gp70, 91, 105, 121
  vascular disease, 88

Immune system function, 2–11
  compromises in, see Immunodeficiencies/immunosuppression
  environmental factors affecting, 11–14; see also Immunodeficiencies/immunosuppression
  hormones and, 10
  immunodeficiency and, 10–11
  and infectious agents, 12–14
  and lymphomagenesis, 57
  and noninfectious agents, 11–12
  in nursing mice, 50
  see also Adaptive immune system/mechanisms
Immunodeficiencies/immunosuppression
  and activation of latent infections, 143–144
  by alkylating agents, 142
  biological inducers, 145–146
  chemically induced, 140–143
  classification of disorders, 11
  defined, 2, 10, 224
  drug-induced, 12, 13; see also Antibiotics; and specific drugs
  environmental causes, 10–11
  genetic causes, 10–11; see also Guinea pig mutants; Hamster mutants; Mouse mutants; Mutations; Rat mutants
  by glucocorticoids, 142–143
  by heavy metals, 11, 141
  by infectious agents, 12–14, 143–144; see also specific agents
  by ionizing radiation, 144–145
  and lymphoid malignancies, model for, 79
  by noninfectious agents, 11–12, 140–143, 144–147
  nutrition and, 144
  by phenols, 141–142
  stress and, 11–12, 50, 146
  by thymectomy, 146–147
  by ultraviolet radiation, 145
  X-linked, 85–87, 90–94
Immunoglobulins
  absence of, 84–85
  B cell production of, 3, 5, 7
  decreased levels of, 47, 70–71, 77, 85–86, 90, 94, 117, 138, 141
  defect in expression of, 63, 86
  defect in isotype switching, 136
  function of, 5–6, 224
  IgA, 50, 61, 70, 84–85, 88, 115, 126
  IgD, 86, 126
  IgE, 117
  IgG, 5, 10, 47, 50, 59, 61, 63, 66, 70, 80, 88, 90, 91, 94, 106, 111, 113, 115, 118, 126, 134, 135, 136, 138, 141, 145, 146
  IgM, 47, 49, 61, 66, 70, 86, 90, 91, 94, 105–107, 110–111, 113, 115, 135, 136, 145
  increased levels of, 58, 61, 66, 91, 94, 105, 107, 110–111; see also Hyperimmunoglobulinemia
  kidney deposits of, 50, 66, 94, 115
  λ1, 118
  and reticulum cell sarcomas, 118
  rheumatoid factors, 61
  species differences in, 5
  structure of, 4–6
  suppression of, 12
Immunologic specificity, see Adaptive immune system/mechanisms
Inbred mouse strains (see also Table 2-1, p. 38)
  129/J, 43, 44
  A/HeJ, 62
  A/J, 39, 119
  AKR, 59, 61, 64, 69, 71, 98, 115–116
  BALB/c, 39, 47, 56, 58, 59, 69, 77, 80, 109, 162, 163
  BRVR, 94
  BRVS, 95
  BXD recombinant, 96
  BXSB, 87–88, 91, 93, 96–98
  CBA/HN, 85, 86
  CBA/H-T6, 68
  CTS, 104
  C3H, 98, 100, 116, 163
  C3HeB/FeJLe, 64, 65
  C3H/HeJ, 17, 34, 39, 40, 42–43, 49, 52–54, 59, 61–64, 86, 88, 90, 157–159
  C3H/HeN, 17, 49, 64, 86
  C57BL, 82
  C57BL/Ka, 77
  C57BL/Ks, 17, 43–45, 72–73, 157
  C57BL/6, 17, 34, 39, 40–45, 49, 52, 59, 61, 63–65, 66, 67, 68, 72–73, 75, 80–84, 87, 88, 90–91, 96, 98, 119, 157–159, 162
  C57BL/10, 109
  C57BL/10J, 100
  C57BL/10ScCr, 62

INDEX 239

C57BL/10ScN, 62
C57BL/10ScSn, 39
C57BL/10Sn, 59
C57/L, 39
DBA/1, 39
DBA/2, 39, 55, 64, 80, 98, 114
DW/J, 43, 49, 156
GL, 51, 157
HRS/J, 56–57, 84, 156
ICR/Jcl, 100
LG/J, 98–99
MRL/Mp, 38, 59, 60–62, 64, 91, 98–100, 121
MWT/Le, 82
NFR/N, 46, 49, 82
NFS/N, 64, 82
N:NIHS, 91
nomenclature for, 17, 34
NZB (New Zealand black), 64, 87, 112, 113, 122
NZW, 105
P/N, 64
RHJ/Le, 56
SB/Le, 42, 87, 93, 96, 97
SJL/J, 52, 59, 61, 62, 64, 87–88
SWR, 112
WB/Re, 82, 83
see also Reproduction
Inbred rats (see also Table 2-1, p. 39)
BBZ/Wor, 133
BN, 161
DA, 126
F344, 126, 163
LEW, 126, 161, 163
PVG, 126
WAG, 126
see also Reproduction
Infectious agents
clinically silent, 12–13
depression of oncogenic virus genes, 13
effects of glucocorticoids on susceptibility to, 143
genetic mechanisms in resistance or susceptibility to, 161–164
immunosuppression by, 12–14
synergistic effects between, 13
X-linked resistance to, 163
see also specific agents
Inflammatory response, 4, 9, 10, 221
Influenza A, 161
Insulin, 10, 14, 17, 43–45, 72, 100, 103, 130–134

Insulinopenia, 72
Insulitis, 100–101, 104, 131, 133
Interferons
activation of NK cells with, 42, 119
defined, 224
γ-, 9, 98, 111, 224
poly(A-U), 145–146, 228
rat beta, 134
and resistance to infectious agents, 163
suppression by environmental agents, 11, 13
type I (α/β), 63, 112
Interleukin (IL)
diminished production of, 91
IL-1, 102, 223
IL-2, 9, 91, 102, 111, 143, 223, 225
IL-3, 223
IL-4, 9, 224
IL-5, 9
IL-6, 9, 98, 224
role in immune response, 9
Islet cells, defects in, 41, 73
Islets of Langerhans
atrophy, 44, 72
degeneration of, 17
destruction of, 131
function of, 224
hypertrophy of, 17
transitional hyperplasia, 101
see also β cells
Ixodid tick, 137

J

Japanese B encephalitis virus, 147

K

K virus, 13, 143
Ketonuria, 101
Kupffer cells, 8

L

Lactic dehydrogenase virus (LDV), 12, 13, 161
Langerhans' cells
diminished, 81
function of, 3, 8, 225
response to ultraviolet radiation, 145
Lead, immunosuppressive effects of, 11, 141

Leishmania donovani, resistance/
  susceptibility mutants, 36, 37, 39–40
Leishmania major, survival mechanism,
  40
Lentiviruses, 13
Leukocytes
  defects in, 41, 74, 76, 84, 126
  peritoneal exudate, 41
Leukocytopenia, 58
Life span, short
  in mice, 65, 115–116
  in rats, 126
  see also Mutations, lethal
Lipopolysaccharide (LPS)
  defined, 225
  enhanced response to, 141
  impaired response to, 17, 37, 62–64, 76,
    84, 94, 96, 137
  macrophage activation by, 95
Listeria monocytogenes, susceptibility to,
  57, 88, 127
Liver parenchymal cells, defects in, 41
Louping ill virus, 94
Lupus, discoid, 135
Lupus nephritis, 108, 113; *see also*
  Systemic lupus erythematosus
Lymph node abnormalities
  enlargement, 47, 53–54, 59–60, 66,
    87–88, 90, 91, 94
  in folicle development, 84
  hyperplasia, 84, 94, 97, 105, 110, 114
  paracortical areas, 49, 126, 144
  small, 50, 58, 77
Lymphadenectomy, 147
Lymphadenopathy, 47, 53–54, 59–60, 66,
  87–88, 90, 91, 94
Lymphocyte abnormalities
  elevation of circulating lymphocytes, 47
  decreased number (lymphopenia), 46, 50,
    66, 72, 74, 77, 78, 110, 126, 131,
    132, 225
  diet-induced, 11
  drug-induced, 12
  in IL-2 production, 54
  lymph node, 46, 77, 126, 138
  peripheral blood, 50, 227
  self-directed cytotoxic lymphocytes, 61
  splenic, 52, 68, 77, 84
  thymic, 84, 126
Lymphocytes
  differentiation antigens, *see* Cell surface
    antigens
  follicular aggregates of, 105
  peripheral blood, 50, 227
  radiosensitivity, 145
  receptors, 10, 60
  repopulation of, 79
  thymic-derived, 54
  see also specific lymphocytes
Lymphocytic choriomeningitis virus
  (LCMV), 12, 13, 14, 147
Lymphokine-activated killer cells,
  independence from NK cells, 93
Lymphokines
  depressed production of, 45
  functions of, 7, 225
  see also Interleukin
Lymphomagenesis, 57, 71
Lymphomas
  autoimmune disease and, 106, 110,
    114
  defined, 225
  follicle center cell, 102
  genes affecting, 117
  immunoblastic sarcoma, 138
  nonthymic, 121
  reticulum cell sarcomas, 71, 117–119,
    228
  thymic, 57, 78
Lymphopoietic cells, abnormality in, 83
Lymphoproliferation, 37, 52–54, 59–62,
  90–91, 105, 110, 141
Lymphosarcoma, 92
Lysosomal function, defects in, 42, 43
Lysosomal granules, giant, 41

M

Macrophage defects, 41, 42
  alveolar, 65, 141
  chemotactic, 68
  in control of pathogen replication, 40
  diet-induced, 11
  function of, 225
  LPS-induced, 63, 64
  in mobilization and listericidal activity, 57
  peritoneal, 68, 76, 129, 143
  and resistance to tolerance induction, 80
  splenic, 40, 68
  in tumoricidal capacity, 63
Macrophages
  alveolar, 8
  endogenously activated, 119, 120
  functions of, 7, 8

monocyte, 3
Maintenance of immunodeficient rodents, see Husbandry practices
Major histocompatibility complex (MHC) antigens
  and C5 deficiency, 55
  Class I, 7, 9, 86, 105, 130, 219
  Class II (Ia), 7–8, 9, 70, 118, 130, 142, 163, 220, 223
  defect in expression of, 63, 105
  and diabetes, 100–101, 130–131
  expression of, 8
  genetic locus, 114, 117, 130–131
  role of, 7, 226
  structure of, 7–8
  T-cell receptor allele associated with responsiveness to, 118
Mammary tumor viruses, 13
Mast cells
  antibody interaction with, 3, 10
  cytoplasmic marker for, 42
  defects in, 41, 83
  function of, 8, 10, 225–226
Mating systems, see Reproduction
Measles, 161
Melanocytes, 81, 83, 226
Melanoma, 73, 92
Membrane attack complex, 55, 226
Mice
  monoclonal antibodies reacting to cell surface antigens in, 213–215
  strains, see Congenic mouse strains; Hybrid mouse strains; Inbred mouse strains; Outbred mouse strains
  stress in, 12
  transgenic, 101
  see also Mouse mutants
Microphthalmia, 37, 51–52, 67–69, 73–77
Microwave irradiation, 145
Mink, Aleutian trait, 41
Minute virus of mice (MVM), 12
Mixed lymphocyte reactions (MLRs), 54, 104, 127, 145, 226
Moloney virus, 13
Monoclonal antibody Mel-14, 60
Mouse adenoviruses, 143
Mouse chromosomes, 6, 7, 18–31, 39, 40, 43, 46, 47, 49, 51, 52, 54, 55, 58, 62, 64, 68, 69, 72, 73, 75, 77, 82, 84, 85–89, 98, 100, 105, 112, 157, 162, 164
Mouse hepatitis virus (MHV), 12, 13, 96, 142, 143, 148, 162, 226

Mouse mutants (see also Table 2-1, pp. 37–38)
  ad (adipose), 43
  A^w (white-bellied agouti), 42
  b (brown), 43
  Bcg^r, Bcg^s (resistance/susceptibility to Mycobacterium bovis), 156
  bg (beige), 15, 92–94, 96–97, 117, 156
  c^e (extreme dilution), 51
  cn (achondroplasia), 99
  db (diabetes), 17, 72, 101, 157
  Dh (dominant hemimelia), 93, 160
  dl^J (downless-J), 157
  dw (dwarf), 75, 156
  ep (pale ears), 42, 43
  gl (grey-lethal), 68, 157
  gld (generalized lymphoproliferative disease), 90, 156
  Hc^0 (hemolytic complement absent), 156
  hr (hairless), 156
  Ity^r, Ity^s (resistance/susceptibility to Salmonella typhimurium), 95–96, 156
  le (light ear), 42
  lh (lethargic), 160
  lpr (lymphoproliferation), 52, 88, 99, 117, 156, 158
  Lps^d (lipopolysaccharide response, defective), 96, 156
  Lsh^r, Lsh^s (resistance/susceptibility to Leishmania donovani), 36, 37, 39–40, 156
  m (misty), 45, 157
  me (motheaten), me^v (viable motheaten), 159
  mi (microphthalmia), 51, 76, 159
  nu (nude), 1, 14, 57, 86, 95, 102, 107, 119–120, 126, 147, 151–152
  $nu\ xid$, 86
  ob (obese), 17, 101, 159
  oc (osteosclerotic), 51, 68, 160
  op (osteopetrosis), 51, 68, 160
  pa (palid), 42, 43
  pe (pearl), 42
  rp (reduced pigmentation), 43
  ru (ruby eye), 42
  ru-2^{mr} (maroon), 42
  sa (satin), 42, 96–97
  $sa\ bg$, 96–97
  $scid$ (severe combined immune deficiency), 14, 45, 156
  Sl (steel), 120

slt (slate), 40
Tol-1 (tolerance to gamma-globulin), 156
W^v (viable dominant spotting), 157
wi (whirler), 43
wst (wasted), 160
xid (X-linked immune deficiency), 54, 156
Yaa (Y-linked autoimmune accelerator), 97
Mouse thymic virus, 143
Mucosa-associated lymphoid tissue (MALT), 8, 225
Multinucleated giant cells, 53
Murine cytomegalovirus, 12
Murine leukemia viruses (MuLVs), 13, 57, 71, 78, 92, 115, 117, 119–120, 121, 222, 226
Murine pinworm (*Syphacia obvelata*), 14
Mutations
  advantages and disadvantages of, 15
  allelic, 1, 36, 37, 39–40, 42, 43, 54–58, 62–71, 82, 125–126
  antigen-selected somatic, 110
  codominant/semidominant, 15, 67, 82, 220, 229
  definition, 15, 226
  lethal, 51–52, 84, 124–125, 156, 158, 159
  maintenance of, 15–16; *see also* Reproduction
  modification of, 16
  multiple, 1, 16, 44–45, 90–94
  nomenclature for, 15, 17, 34
  provirus integration into normal allele, 56
  radiation-induced, 40, 42, 67
  sex-linked, 85
  single point, 1, 36–90
  spontaneous, 15, 40, 43, 47, 51, 52, 58, 59, 64, 69, 72, 73, 75, 77, 80, 84, 123, 124, 128
  see also Hamster mutants; Mouse mutants; Rat mutants
Mycobacterium bovis, resistance/susceptibility mutants, 36, 37, 39–40, 127
Mycobacterium lepraemurium, resistance/susceptibility mutants, 36, 39–40
Mycoplasma pulmonis, 128, 163
Myocardial infarction, 60, 110

N

Natural killer (NK) cells
  activation of, 42

  depression of, 42, 43, 91–94, 102, 116, 119–120, 142, 146
  increased activity, 70, 71, 73, 93, 126
  independence from lymphokine-activated killer cells, 93
  lack of, 104
  role of, 10, 226
  serotonin dependency of, 10
  source of, 10
  and tumor growth, 24, 92
Nematode infestation, resistance to, 83, 164
Neurological disorders, mouse mutants with, 42–43, 84
Neutrophils, *see* Polymorphonuclear leukocytes
Nippostrongylus brasiliensis, 83
Nocardia (mitogen), diminished response to, 86
Nomenclature, 15, 17, 34
Nutrition, and immunosuppression, 144; *see also* Diet

O

Obesity, 17, 43–45, 72–73, 104
Oligodactyly, 47
Oncogenic virus genes, suppression of, 13
Oncornaviruses, immunosuppression by, 13, 227; *see also specific viruses*
Osteoclasts
  cytoplasmic marker for, 42
  defects in, 51, 68, 74, 76, 123, 124–125, 129
  elevated acid phosphatase, 123
  function of, 227
  reduced acid phosphatase in, 51, 68, 76
Osteopetrosis
  defined, 227
  in mice, 51–52, 67–69, 73–77
  in rats, 123–125
Osteoporosis, 116
Osteosclerosis, 73–74, 227
Outbred mice (*see also* Table 2-1, p. 39)
  sensitivity to ectromelia virus, 162

P

Pancreatic lymphocyte infiltration, 131
Paracoccidioidomycosis, 163
Parasites, delayed expulsion of, 83
Paroviruses, 143
Pericytes, cerebral cortex, defects in, 41

INDEX   243

Periodic acid-Schiff (PAS)-positive cells, 50
Peyer's patches
   absence of IgA-specific B-cell precursors from, 85
   atrophy of, 144
   function of, 8, 227
   small, 58, 66, 84, 126
Phytohemagglutinin (PHA)-induced blastogenesis, 46, 227
Picryl chloride, contact sensitivity to, 81
Plaque-forming cell (PFC) responses, impaired, 46, 63, 227
Plasma cells, with Russell bodies, 66
Pneumococcal polysaccharide, failure to respond to, 85
Pneumocystis carinii, 79, 128, 142, 143, 153
Pneumocytes, type II, defects in, 41
Pneumonitis
   interstitial, 54, 60, 79
   spontaneous, susceptibility to, 42, 128
Polyarteritis, 60, 99, 121
Polybrominated biphenyls, 140
Polydactyly, 47, 48
Polydipsia, 101, 131
Poly(A-U), 145–146, 228
Poly(I-C), 119, 228
Polymorphonuclear leukocytes (PMNs)
   aggregates in skin, 65
   antibody interaction with, 3
   basophils, 10
   cytoplasmic marker for, 42
   decreased number of, 126
   function of, 8, 228
   impaired chemotactic and bactericidal activities, 41–42, 141
   increased number of, 47, 66
   response to LPS, 63
   see also Mast cells
Polyomavirus, 147, 161
Polyphagia, 101
Polyunsaturated fatty acids, 144
Polyuria, 101
Progenitor cell replication, model for cell-associated molecules limiting, 64–66
Prokaryotic cells, 4
Prolactin, 10, 46
Proliferative enteritis, 139
Propylgallate, 141–142, 227
Proteinuria, 91, 99, 110
Proximal tubule cells, defects in, 41
Pseudomonas aeruginosa, 144–145
Purkinje cells, cerebellar, defects in, 41, 43, 84
Pyogenic bacteria, susceptibility of mutants to, 42
Pyramidal cells, defects in, 41

R

Radiation, ionizing, and immunosuppression, 144–145
Rat mutants (*see also* Table 2-1, p. 39)
   fa (fatty), 124, 133
   op (osteopetrosis), 123
   rnu (Rowett nude), 1, 14, 147
   tl (toothless), 39, 123, 124, 128–129
Rats
   adjuvant arthritis in, 14
   immunoglobulins, 5
   linkage map, 32–33
   monoclonal antibodies reacting to cell surface antigens in, 213–215
   stress in, 12
   see also Inbred rats; Rat mutants
Raucher virus, 13
Renal disease, 50, 54, 61, 66, 91, 94, 99, 108, 110, 112, 114, 121
Reovirus 3, 12, 14
Reproduction
   backcross linkage analysis, 63
   backcrosses, 59, 64, 86–88, 99, 100–101, 126, 156–158, 160, 218
   balanced stock, 156–157, 219
   brother × sister mating, 52, 87, 103, 121, 139, 155–156, 158, 159
   cross–intercross cycles, 52, 59, 158–159, 160
   diabetic mice, 45, 103, 157
   double heterozygote repulsion mating, 157
   heterozygous matings, 45, 47, 51, 67, 69, 73, 76, 83–85, 125, 128, 129, 156, 158
   high-performance, 100
   homozygous, 125, 156, 158
   homozygous males and heterozygous females, 58, 71, 124, 128, 139
   hybrid matings, 99
   inbreeding, 62, 155–159
   without inbreeding, 159–160
   intercrossing, 156–158, 224
   outcrossing, 100

ovarian transplantation method, 67, 73, 77, 85, 158–159, 160, 227
paired matings, 89, 98
random matings, 122, 123, 124
after thymus transplants, 125
transferring a mutation to a hybrid background, 159–160
transferring a mutation to an inbred background, 59, 64, 86–88, 99, 100–101, 126, 156, 157–159
trio matings, 62
Reticuloendothelial system
function of, 228
reduced function in, 88
stimulant, 146
Reticulum cell sarcomas, 71, 117–119, 228
Retroviral genes, 44, 107
Retroviruses, see Murine leukemia viruses; Oncornaviruses
Rickets, 74
Rickettsia akari, 163
Rickettsia tsutsugamushi, 163

S

Salmonella choleraesuis, 137
Salmonella typhi, 40
Salmonella typhimurium
absence of *aroA* gene, 96
impaired clearance from mucosal surfaces, 63
index of killing, 96
resistance to, 36, 37, 39–40, 96
susceptibility to, 36, 37, 39–40, 62–63, 95–96, 143
Sendai virus, 12, 13, 121, 128, 143, 148, 161, 162
Senescence, early, 115–117
Serotonin
and bactericidal activity of PMNs, 42
decreased levels in platelets, 42
function of, 10, 229
Serum amyloid A, 63, 229
Severe combined immunodeficiency, 14, 37, 45, 77–80, 85, 92
Sheep red blood cells (SRBCs)
defined, 229
reduced immune response to, 46, 47, 49, 81, 84, 115, 116, 121, 141–142, 146
Shock, complement system activation and, 9–10
Skin allograft rejection

impaired ability for, 46, 58, 110, 119, 132
infectious agents and, 13
Spinal cord neurons, defects in, 41
Spleen, model of functional relationships with thymus and bone marrow, 93
Spleen transplantation
correction of defects by, 68–69
transfer of syngeneic spleen cells, 146
Splenectomy, 147
Splenic abnormalities
absence of B colonies, 94
asplenia, 47, 93
decreased lymphoid cells, 50
hyperplasia, 106, 110
periarteriolar sheaths, 126
perifollicular sheaths, 49
reduced cellularity, 51
small organ, 58, 77
splenomegaly, 54, 60, 65–66, 68, 87–88, 94, 97, 106, 229
stem cell increases, 68, 76
in T cells, 57, 125
ThB$^+$ cells, 94
St. Louis encephalitis virus, 94
Staphylococcus aureus, delayed clearance of, 55
Starvation, 51
Stem cells
depression of, 141
differentiation, 83
function of, 229
hematopoietic, 3, 141
radiosensitivity, 145
splenic, 51, 68, 76, 141
Sterility, 66, 82, 156, 158
Streptococcal group A carbohydrate (GAC), impaired response to, 95
Stress, and immunosuppression, 11–12, 142, 146
Submandibular gland, leukocyte infiltrates of, 104
Systemic lupus erythematosus, 54, 99, 105, 108–111, 114–115, 121, 122, 135, 229

T

T cell defects
absence of mature cells, 92–94
antibody response against erythrocytes, 107

β-chain, deletion in, 109
bone marrow TdT$^+$ cells, 66
decreased number of cells, 46, 49, 66, 70–71, 126–127, 131, 132, 145
deletion of V$_\beta$T-cell receptor genes, 118
hyperplasia, 91
increased number of cells, 47, 101
Ly-1$^+$ expansion, 91
in Ly-5 expression, 60
in Ly-123$^+$ subset, 107
in lymph nodes, 53–54
lymphopenia, 45, 104, 145
in mitogen responses, 104, 125, 141
NTA-reactive antigen, 106
production of pathogenic IgG anti-DNA, 109
proliferation abnormality, 54, 57, 66, 78
response to I-region alloantigens, 57
splenic, 57
suppressor-inducer (Ts), 50, 102, 106, 119, 121, 144
Tc, 41, 42, 61, 66, 132
Th, 47, 57, 70, 112–113, 116, 119, 132, 141
in Thy-1$^+$ populations, 63, 110, 113
T cells
activation of, 7, 9, 70
CD4, 113
CD8, 113, 142
cytotoxic (Tc), 7, 9, 10, 41, 42, 70, 220; see also Natural killer cells
differentiation, 7, see Cell surface antigens
estrogen receptors, 10
function of, 3, 8, 9, 229–230
growth factors, 9; see also Interleukin
helper-inducer (Th), 7, 9, 47, 57, 91, 142, 230
Ia antigen expression on, 8, 9
interaction with B cells, 3, 7, 9, 91, 92, 109, 112, 116, 119
macrophage activation by, 95
model for thymus-independent maturation of precursors to functional effector cells, 127
proliferation of, 3, 7, 54
structure, 7
surface-membrane inserted antigen receptor (Ti), 7, 10
Ts, 9, 230
see also Lymphocytes
Taenia teaniaformis, 164

Tetrachlorodibenza-*p*-dioxin, 140, 229
Tetracyclines, 12
Theiler's murine encephalomyelitis virus, 117, 163
Thrombocytes, elevation of, 47
Thymectomy, 146–147
Thymic defects
absence of Ia$^+$ cells, 70
athymia, 93, 138, 148
atrophy, 49–50, 57, 110, 142, 144
B-cell influx, 119
epithelial dysgenesis, 70
hyperplasia, 105, 110, 114
involution, 45, 51, 65, 105, 140
lymphomas, 57, 78, 79
reduced cellularity, 84
small organ, 58, 65, 77
Thymulin, 50, 105, 230
Thymus
function of, 230
model of functional relationships with spleen and bone marrow, 93
see also Thymic defects
Thyroid defects
calcitonin-secreting cells, 74, 76
follicular cells, 41, 51
Hashimoto-like thyroiditis, 132, 133, 222
thyroiditis, 95
Thyrotropic hormone-producing cells, deficiencies in, 49
Thyroxine, 10
Tonsils, 8
Transmissible ileal hyperplasia, 139
Trichinella spiralis, 83, 147, 164
Trichuris muris, 164
Tris, 140
Tumor growth and metastasis
anesthetic agents and, 12
models for, 69–71, 92
NK cell activity and, 92
Tumors and tumor cell lines
B16F10, 92
CEM, 92
contaminants in, 12
human, 92
immunosuppression by, 146
impaired cytolysis of, 42
K562, 92
LOX, 92
murine, 92
ovarian granulosa cell tumors, 92
reticulum cell neoplasms, 99

suppression of, 13
YAC-1, 92
Tyzzer's disease, 128

U

Ultraviolet radiation, and immunosuppression, 145, 230–231

V

Variable-region-coding elements, aberrant deletions of, 78

Vascular disease, 97, 110
Vasculitis, 135, 114, 222

W

Wasting disease, 71, 84–85
Water, hyperchlorinated acidified, 11, 152–153

X

Xenografts, 79, 231